T0324853

An Elementary Introduction to Statistical Learning Theory

An Elementary Introduction to Statistical Learning Theory

SANJEEV KULKARNI

Department of Electrical Engineering
School of Engineering and Applied Science
Princeton University
Princeton, New Jersey

GILBERT HARMAN

Department of Philosophy
Princeton University
Princeton, New Jersey

A JOHN WILEY & SONS, INC., PUBLICATION

Published by John Wiley & Sons, Inc., Hoboken, New Jersey
Published simultaneously in Canada

For general information on our other products and services or for technical support, please contact our
Customer Care Department within the United States at (800) 762-2974, outside the United States at
(317) 572-3993 or fax (317) 572-4002.

Wiley publishes in a variety of print and electronic formats and by print-on-demand. Some material
included with standard print versions of this book may not be included in e-books or in print-on-demand.
If this book refers to media such as a CD or DVD that is not included in the version you purchased,
you may download this material at http://booksupport.wiley.com. For more information about Wiley
products, visit www.wiley.com

Library of Congress Cataloging-in-Publication Data:
Kulkarni, Sanjeev.
    An elementary introduction to statistical learning theory / Sanjeev Kulkarni, Gilbert Harman.
        p. cm.—(Wiley series in probability and statistics)
    Includes index.
    ISBN 978-0-470-64183-5 (cloth)
  1. Machine learning–Statistical methods. 2. Pattern recognition systems. I. Harman, Gilbert. II. Title.
    Q325.5.K85 2011
    006.3′1–dc22

                                                                    2010045223

Printed in Singapore

oBook ISBN: 978-1-118-02347-1
ePDF ISBN: 978-1-118-02343-3
ePub ISBN: 978-1-118-02346-4

10 9 8 7 6 5 4 3 2 1

Contents

Preface

This book offers a broad and accessible introduction to the relatively new field of statistical learning theory, a field that has emerged from engineering studies of pattern recognition and machine learning, developments in nonparametric statistics, computer science, the study of language learning in linguistics, developmental and cognitive psychology, the philosophical problem of induction, and the philosophy of science and method.

The book is the product of a very successful introductory course on "Learning Theory and Epistemology" that we have been teaching jointly in electrical engineering and philosophy at Princeton University. The course is open to all students and has no specific prerequisites other than some analytical skills and intellectual curiosity. Although much of the material is technical, we have found that the main points are both accessible to and appreciated by a broad range of students. In each class, our students have included freshmen through seniors, with majors from the sciences, engineering, humanities, and social sciences.

The engineering study of pattern recognition is concerned with developing automated systems to discriminate between various inputs in a useful way. How can the post office develop systems to scan and sort mail on the basis of hand-written addresses? How can a manufacturer design a computerized system to transcribe ordinary conversations? Can computers be used to analyze medical images to make diagnoses?

Machine learning provides an efficient way to approach some pattern recognition problems. It is possible to train a system to recognize handwritten zip codes. Automated systems can interact with users to learn to perform speech recognition. A computer might use machine learning to develop a system that can analyze medical images in the way that experts do.

Machine learning and pattern recognition are also concerned with the general principles involved in learning systems. Rather than develop algorithms from scratch and in an ad hoc manner for each new application, a systematic methodology can be extremely useful. It is also important to have techniques for evaluating the performance of a learning system. Knowing what is achievable and what is not helps to provide a benchmark and often suggests new techniques for practical learning algorithms.

These questions are also related to philosophical questions that arise in epistemology. What can we learn and how can we learn it? What can we learn about other minds and the external world? What can we learn through induction?

The philosophical problem of induction asks how it is possible to learn anything on the basis of inductive reasoning, given that the truth of the premises of inductive reasoning does not guarantee the truth of its conclusion. There is no single solution to this problem, not because there is no solution, but because there are many, depending on what counts as learning. In this book, we explain how various solutions depend on the way the problem of induction is formulated.

Thus, we hope this book will serve as an accessible introduction to statistical learning theory for a broad audience. For those interested in more in-depth studies of learning theory or practical algorithms, we hope the book will provide a helpful starting point. For those interested in epistemology or philosophy in general, we hope the book will help draw connections into very relevant ideas from other fields. And for others, we hope the book will help provide an understanding of some deep and fundamental insights from statistical learning theory that are at the heart of advances in artificial intelligence and shed light on the nature and limits of learning.

We acknowledge with thanks a Curriculum Development Grant from the 250th Anniversary Fund for Innovation in Undergraduate Education from Princeton University. Rajeev Kulkarni gave us extremely useful comments on the whole book, which has greatly improved the result. Joel Predd and Maya Gupta also provided valuable comments on various parts. We have also benefitted from a careful reading by Joshua Harris. We are also grateful to our teaching assistants over the years and to the many students who have discussed the content of the course with us. Thanks!

CHAPTER 1

Introduction: Classification, Learning, Features, and Applications

1.1 SCOPE

In this book we are concerned mainly with pattern classification—classifying an object into one of several categories on the basis of several observations or measurements of the object. The simplest case is classification of an object into one of two categories, but a more general case allows for any finite number of categories.

A second closely related task is estimation of a real number that is typically related to some property of the object. As in classification, several observations or measurements of the object are available, and our estimate is based on these observations.

Most of our discussion concerns issues arising about the first task, classification. But we occasionally say something about the second task, estimation. In either case, we are interested in *rules* for classifying objects or estimating values, given certain observations or measurements. More specifically, we are interested in methods for *learning* rules for classification or estimation.

We discuss some concrete examples further below. For now, think about learning to recognize handwritten characters or faces or other objects from visual data. Or, think about the problem of recognizing spoken words. While humans are extremely good at these types of classification problems in many natural settings, it is quite difficult to design automated algorithms for these tasks with performance and robustness anywhere near those of humans.

Even after more than a half century of effort in fields such as electrical engineering, mathematics, computer science, statistics, philosophy, and cognitive science, humans can still far outperform the best machine learning algorithms that have ever been developed. That said, enormous progress has been made in learning theory,

An Elementary Introduction to Statistical Learning Theory, First Edition.
Sanjeev Kulkarni and Gilbert Harman.
© 2011 John Wiley & Sons, Inc. Published 2011 by John Wiley & Sons, Inc.

algorithms, and applications. Results in this area are deep and practical and are relevant to a range of disciplines such as those we have mentioned above. Many of the basic ideas are accessible to a broad audience. However, most treatments of this material are at an advanced level, requiring a rather technical background and expertise.

Our aim in this book is to provide an accessible introduction to this field, either as a first step for those wishing to pursue the subject in more depth, or for those desiring a broad understanding of the basic ideas. For most of the book, we focus on the problem of two-class pattern classification. This problem arises in many useful applications and is sufficiently rich to explain many of the key ideas in the field, yet removes some unnecessary complications. Although many important aspects of learning are not covered by this model, we provide many good references for more depth, generalizations, and other models. We hope this book will serve as a valuable entry point.

1.2 WHY MACHINE LEARNING?

Algorithms for recognizing patterns would be useful in a wide range of problems. This ability is one aspect of "artificial intelligence." But one might reasonably ask why we need to design automated methods for *learning* good rules for classification, as opposed to just figuring out what is a good rule for a given application and implementing it.

The main reason is that in many applications, the only way we can find a good rule is to use data to learn one. For example, it is very hard to describe exactly what constitutes a face in an image, and therefore it is hard to come up with a classification rule to decide whether or not a given image contains a face. But, given a good learning algorithm, we might be able to present the algorithm with many examples of images of a face and many examples of images without a face, and then let the algorithm come up with a good rule for recognizing whether or not a face is present. There are other benefits of having a learning algorithms as well, such as robustness to errors in assumptions or modelling, reduced need for explicit programming, and adaptation to changing conditions.

In general, for a classification problem, we want to decide to which of several categories the object belongs on the basis of some measurements of the object. To *learn* a good rule, we use data that consist of many examples of objects with their correct classification. The following questions immediately arise:

1. What do we mean by an "object" and "measurements" of the object?
2. In the classification problem, what are the categories to which we assign objects?
3. In the estimation problem, what are the values we attempt to estimate?
4. How do we measure the quality of a classification or estimation rule, and what is the best rule we could hope for?

5. What information is available to use for learning?

6. How do we go about learning a good classification or an estimation rule?

We describe the answers to the first three questions in this chapter. To answer the remaining questions, some background material on probability is provided in Chapters 2 and 3. With this background, the answer to the fourth question is discussed in Chapters 4 and 5. The answer to the fifth question is discussed in Chapter 6. The rest of the book is devoted to various aspects of and approaches to the last question.

1.3 SOME APPLICATIONS

Before discussing further details, it may be helpful to have some concrete examples in mind. There are a wide range of applications for learning, classification, and estimation. Here we mention just a few.

1.3.1 Image Recognition

There are many applications in which the object to be classified is a digital image. The "measurements" in this case might describe the outputs of each of the pixels in the image. In the case of a black and white image, the intensity of each pixel serves as one measurement. If the image has $N \times N$ pixels, then the total number of pixels (and hence measurements) is N^2. In the case of a color image, each pixel can be considered as providing three measurements, corresponding to the intensities of each of three color components, say RGB values. Hence, for an $N \times N$ color image, there are $3N^2$ measurements.

Depending on the application, there are many classification tasks based on using these measurements. Face detection or recognition is a common and useful application. In this case, the "categories" might be face versus no face present, or there might be a separate category for each person in a database of individuals.

A different application is character recognition. In this case, the writing can be segmented into smaller images that each contain a single character, and the *categories* might consist of the 26 letters of the alphabet (52 letters, if upper and lower case letters are to be distinguished), the 10 digits, and possibly some special characters (period, question mark, comma, colon, etc.).

In yet another application, the images might be of industrial parts and the categorization task is to decide whether the current part is defective or not.

1.3.2 Speech Recognition

In speech recognition, we are interested in recognizing the words uttered by a speaker. The *measurements* in this application might be a set of numbers that represent the speech signal. First, the signal is typically segmented into portions that contain distinct words or phonemes. In each segment, the speech signal can be represented in a variety of ways. For example, the signal can be represented by the intensities or energy in different time-frequency bands. Although the details of the signal representation are outside the scope of this book, the signal can be represented ultimately by a set of real values.

In the simplest case, the *categories* might be as simple as deciding whether the utterance is "yes" versus "no." A slightly more complicated task might be to decide which of the 10 digits is uttered. Or there might be a category for each word from a large dictionary of acceptable words and the task might be to decide which, if any, of this large number of words has been uttered.

1.3.3 Medical Diagnosis

In medical diagnosis, we are interested in whether or not there is a disease present (and which disease). There is a separate *category* for each of the diseases under consideration and one category for the case where no disease is present.

The *measurements* in this application are typically the results of certain medical tests (e.g., blood pressure, temperature, and various blood tests) or medical diagnostics (such as medical images), presence/absence/intensity of various symptoms, and some basic physical information about the patient (age, sex, weight, etc.).

On the basis of the results of the measurements, we would like to decide which disease (if any) is present.

1.3.4 Statistical Arbitrage

In finance, statistical arbitrage refers to automated trading strategies that are typically of a very short term and involve a large number of securities. In such strategies, one tries to design a trading algorithm for the set of securities on the basis of quantities such as historical correlations among the large set of securities, price movements over recent time horizons, and general economic/financial variables. These can be thought of as the "measurements" and the prediction can be cast as a classification or estimation problem. In the case of classification, the categories might be "buy," "sell," or "do nothing" for each security. In the estimation case, one might try to predict the expected return of each security over some future time horizon. In this case, one typically needs to use the estimates of the expected return to make a trading decision (buy, sell, etc.).

1.4 MEASUREMENTS, FEATURES, AND FEATURE VECTORS

As we discussed in Sections 1.1 and 1.3, in classifying an object, we use observations about the object in order to make our decision. For example, when humans wish to classify an object, they might look at the object, pick it up, feel it, listen to it, etc. Or they might use some instruments to measure other properties of the object such as size, weight, and temperature.

Similarly, when designing a machine to automatically classify (or learn to classify) objects, we assume that the machine has access to measurements of various properties of the object. These measurements come from sensors that capture some physical variables of interest, or *features*, of the object.

For simplicity, in this book we model each measurement (or feature) as being captured by a single real number. Although in some applications, certain features may not be very naturally represented by a number, this assumption allows discussion of the most common learning techniques that are useful in the most common applications.

We assume that all the relevant and available aspects of the objects can be captured in a finite number of measurements/features. These finite number of features can be put together to form a *feature vector*. Suppose there are d features with the value of the features given by $x_1, x_2, \ldots, x_d$. The feature vector is $\overline{x} = (x_1, x_2, \ldots, x_d)$. This feature vector can be thought of as a point or a vector in d-dimensional space $\mathbf{R}^d$, which we call the *feature space*. Each component of the feature vector, indicating the value of the corresponding feature, is the value along a particular dimension of the feature space.

In the case of image recognition with an $N \times N$ image, the number of features is N^2 for a black and white image and $3N^2$ for a color image.

In speech recognition, the number of features is equal to the number of real values used to represent the speech segment to be classified.

1.5 THE NEED FOR PROBABILITY

In most applications, the category of the object is not uniquely and definitively determined by the value of the feature vector. There are some fundamental reasons for this. First, although it would be nice if the measured features capture all the properties of the object important for classification, this is usually not the case. The measured features might fail to capture some important details. This should be clear in the examples given above.

Second, depending on the application and the specific measurements, the feature values may be noisy. That is, there may be some inherent uncertainty or randomness in the observed values of the features so that even the same object might give rise to different values on different occasions.

For these reasons, it is helpful to use tools from probability to formulate the problem precisely and guide the solution. In Chapters 2 and 3, we review some of the basic tools from probability that we need for the rest of the book.

1.6 SUPERVISED LEARNING

After providing the necessary background from probability, in Chapter 4, we formulate the pattern recognition problem. In the ideal (and unusual) case, where the underlying probabilistic structure is known, the solution to the classification problem is well known and is a basic result from statistics. This is discussed in Chapter 5.

However, in the much more typical case in applications, the underlying probability distributions are not known. In this case, we try to overcome this lack of knowledge by resorting to labeled examples as we discuss in Chapter 6. The learning problem, as formulated in Chapter 6, is just one type of machine learning problem known by various terms such as *learning from examples*, *supervised learning*, *statistical pattern classification*, *statistical pattern recognition*, and *statistical learning*.

The term "supervised" learning arises from the fact that examples we assume that we have access to are properly labeled by a "supervisor" or "teacher." This contrasts with "unsupervised learning," in which many examples of objects are available, but the class to which the objects belong are unknown. There are also other formulations of machine learning problems such as semi-supervised learning and reinforcement learning, as well as many other related problems in statistics, computer science, and other fields. But in this book, we focus exclusively on the case of supervised learning.

1.7 SUMMARY

In this chapter, we described the general problems of classification and estimation and discussed several concrete and important applications. We then introduced the terminology of features, feature vectors, and feature space. The need for introducing probability and learning was described.

We have mentioned both classification and estimation. We focus mainly on classification in this book, with some discussion of extensions to estimation.

In the next two chapters, we review some principles of probability that are important for aspects discussed in the rest of the book. After this, we formalize the classification (or pattern recognition) problem and discuss general issues in learning from data, before moving on to a discussion of specific learning methods and results.

1.8 APPENDIX: INDUCTION

The appendices at the end of each chapter briefly discuss certain side issues, perhaps of a philosophical nature.

In this book, we are concerned primarily with inductive learning rather than deductive learning. Deductive learning consists in deriving a new conclusion from premises whose truth guarantees the truth of the conclusion. For example, you might learn that the area of a parallelogram is equal to its base times its height by deducing this from what you already know about rectangles and about how the area of a parallelogram is related to the area of a rectangle with the same base and height. You might then learn that the area of a triangle is equal to its base times *half* its height, by deducing this from the fact that any triangle is exactly half of a certain parallelogram.

Inductive learning consists in reaching a conclusion from evidence that does not guarantee the truth of the conclusion. For example, you might infer from the fact that mail has almost always been delivered before noon on Saturdays up until now to the conclusion that mail will be delivered before noon next Saturday. This is an inductive inference, because the data do not guarantee the truth of the conclusion. Sometimes, the conclusion of an inductive inference is false even though the "premises" of the inference are all true.

The philosophical "problem of induction" asks how one can be *justified* in believing inductive conclusions from true premises. Certainly, it is not possible to prove deductively that any such inductive conclusion is true if its premises are, since typical inductive inferences do not provide such a guarantee. Even if you are justified inductively in thinking that your mail will be delivered before noon next Saturday, it is compatible with your evidence that your mail is not delivered before noon next Saturday. Induction is not a special case of deduction.

It might be suggested that induction has almost always led to true conclusions in the past, so it is reasonable to conclude that it will almost always lead to true conclusions in the future. The objection to this suggestion is that this is circular reasoning: we are assuming that induction is justified in order to argue that induction is justified!

On the other hand, is it possible to offer a noncircular justification of deduction? Wouldn't any such justification take the form of a deductive argument and so also be circular?

It will emerge that statistical learning theory provides partial deductive mathematical justifications for certain inductive methods, given certain assumptions.

1.9 QUESTIONS

1. What is a feature space? What do the dimensions of such a space represent? What is a vector? What is a feature vector?

2. If we want to use the values of F different features, in order to classify objects, where each feature can have any of G different values, what is the dimension of the feature space?

3. For a 12×12 grayscale image (256 grayscale levels), how many *dimensions* are there for the feature vector? How many different possible *feature vectors* are there?

4. Is classification a special case of estimation? What differences are there between typical cases of classification and typical cases of estimation?

5. About the problem of induction

   (a) What is the problem of induction?
   (b) How does the reliability of induction compare with the reliability of deduction?
   (c) How might statistical learning theory say something about the reliability of induction?

1.10 REFERENCES

Statistical pattern recognition as a distinct field has been an active area of research for about half a century, though its foundations are based on probability and statistics which go back much further than that. Statistical pattern recognition (or statistical learning) is part of the broad area of machine learning and spans many disciplines such as mathematics, probability, statistics, electrical engineering, computer science, cognitive science, econometrics, and philosophy. There are a number of conference, journals, and books devoted to machine learning, and among these much of the material is devoted to statistical learning.

Mitchell (1997) is an introduction to issues about machine learning generally. Vickers (2010) is an up-to-date discussion of the problem of induction. The other references below are just some of the many classic and recent references that discuss statistical pattern recognition and related areas at various levels.

Bishop C. Pattern recognition and machine learning. New York: Springer; 2006.

Bongard M. Pattern recognition. Washington (DC): Spartan Books; 1970.

Devijver PR, Kittler J. Pattern recognition: a statistical approach. Englewood Cliffs (NJ): Prentice-Hall; 1982.

Devroye L, Györfi L, Lugosi G. A probabilistic theory of pattern recognition. New York: Springer Verlag; 1996.

Duda RO, Hart PE. Pattern classification and scene analysis. New York: Wiley; 1973.

Duda RO, Hart PE, Stork DG. Pattern classification. 2nd ed. New York: Wiley; 2001.

Fukunaga K. Introduction to statistical pattern recognition. 2nd ed. San Diego (CA): Academic Press; 1990.

Hastie T, Tibshirani R, Friedman J. The elements of statistical learning: data mining, inference, and prediction. 2nd ed. New York: Springer; 2009.

Ho YC, Agrawala A. On pattern classification algorithms: introduction and survey. Proc IEEE 1968;56:2101–2114.

Kulkarni SR, Lugosi G, Venkatesh S. Learning pattern classification - A survey. IEEE Trans Inf Theory 1998; 44(6): 2178–2206.

Mitchell T. Machine learning. Boston (MA): McGraw-Hill; 1997.

Nilsson NJ. Learning machines. New York: McGraw-Hill; 1965.

Schalkoff RJ. Pattern recognition: statistical, structural, and neural approaches. New York: Wiley; 1992.

Theodoridis S, Koutroumbas K. Pattern recognition. 4th ed. Amsterdam: Academic Press; 2008.

Theodoridis S, Pikrakis A, Koutroumbas K, Cavouras D. Introduction to pattern recognition: a matlab approach. Amsterdam: Academic Press; 2010.

Vapnik VN. The nature of statistical learning theory. New York: Springer; 1999.

Vickers J. The Problem of Induction, in The Stanford Encylopedia of Philosophy; 2010, http://plato.stanford.edu/entries/induction-problem/.

Watanabe MS. Knowing and guessing. New York: Wiley; 1969.

CHAPTER 2

Probability

In this and the next chapter, we explain some of the elementary mathematics of probability. This provides the mathematical foundation for dealing with uncertainty and forms the basis for statistical learning theory. In particular, we are interested in learning when there is uncertainty in the underlying objects (feature vectors), the labels (indicating the class to which the objects belong), and the relationship between the class of the object and the feature vector. This uncertainty will be modeled probabilistically.

In this chapter, we explain some of the basics of discrete probability. That is, we focus on the case with a finite number of outcomes of interest. In the next chapter, we briefly discuss the continuous case, in which probability density becomes important. In the appendix we discuss some possible interpretations of probability.

2.1 PROBABILITY OF SOME BASIC EVENTS

Suppose that we have an opaque bowl containing balls of various colors. A ball is randomly drawn from the bowl, its color is noted, and then it is placed back in the bowl. The probability of getting a ball of a particular color is the fraction of the balls in the bowl that have that color.

The assumption that balls are drawn randomly means that the probability of drawing any particular ball is equal to the probability of drawing any other ball. That is an assumption about one sort of case. We do not assume the general application of the principle of indifference, which says that if all we know is that there are N distinct possibilities, then the probability of any one is equal to the probability of any other. One problem with this principle is that there can be different ways of counting possibilities. For example, if we toss two coins, should we say there are three possibilities: two heads, two tails, and one of each? Or two

An Elementary Introduction to Statistical Learning Theory, First Edition.
Sanjeev Kulkarni and Gilbert Harman.
© 2011 John Wiley & Sons, Inc. Published 2011 by John Wiley & Sons, Inc.

possibilities: the coins are same versus different? Or four possibilities: head/head, head/tail, tail/head, and tail/tail?

Suppose that there are N balls in the bowl, B of them are black, W of them are white, and that none is both black and white. (For example, none of the balls has black and white stripes.) Then the probability of getting a black ball is B/N and the probability of getting a white ball is W/N.

Notation: Let x_n denote the color of the nth ball drawn. So, "x_1 is black" means that the first ball drawn is black and "x_2 is white" means that the second ball drawn is white. For an event S, $P(S)$ denotes the probability of S. So, $P(x_n$ is white) denotes the probability that the nth ball is white.

Then we have $P(x_n$ is black) $= B/N$ and $P(x_n$ is white) $= W/N$.

We are assuming that no ball is both black and white, so we are assuming that it is impossible that the nth ball drawn is both black and white. In our notation, we are assuming that the following is impossible: $(x_n$ is black)$\&(x_n$ is white).

Consider the probability that the nth ball drawn is both black and white, $P((x_n$ is black)$\&(x_n$ is white)). Since none of the balls is both black and white, the fraction of the balls that are both black and white is 0. So, we have $P((x_n$ is black)$\&(x_n$ is white)) $= 0$.

In general: *The probability of an impossible event is 0*.

Given that no balls are both black and white, what is the probability that either a black ball or a white ball is drawn? That probability is the fraction of balls that are either black or white. That fraction is the sum of the fraction of balls that are black plus the fraction of balls that are white, $B/N + W/N = (B + W)/N$. So, using "$\vee$" to mean "or,"

$$P((x_n \text{ is black}) \vee (x_n \text{ is white})) = P(x_n \text{ is black}) + P(x_n \text{ is white}).$$

In general: *The probability that one of several incompatible events will occur is the sum of their probabilities*.

Suppose every ball is either black or white. Then the fraction of balls that are either black or white is $N/N = 1$.

$$P((x_n \text{ is black}) \vee (x_n \text{ is white})) = 1.$$

In general: *An event that is guaranteed to occur has the probability 1*.

It is certain that the nth ball drawn will be either black or not black. Let us use "$\neg(x_n$ is black)" to mean that the ball drawn in the nth drawing is not black. Then we have

$$P((x_n \text{ is black}) \vee \neg(x_n \text{ is black})) = 1.$$

Since a ball cannot be both black and not black, we have

$$P(x_n \text{ is black}) + P(\neg(x_n \text{ is black})) = 1.$$

Subtracting the first term from both sides of the equation yields

$$P(\neg(x_n \text{ is black})) = 1 - P(x_n \text{ is black}).$$

In general: *The probability of any event is 1 minus the probability that the event will not occur.*

$$P(S) = 1 - P(\neg S).$$

2.2 PROBABILITIES OF COMPOUND EVENTS

Suppose that we draw two balls, one after the other, putting back the first ball and thoroughly mixing the balls before drawing the second. (This is called "drawing with replacement.") What is the probability of getting a black ball the first time and a white ball the second time?

We have to calculate how many possibilities there are in all, and how many of those possibilities are ones in which the first ball is black and the second is white. Since we are equally likely to get any of the N balls from the bowl the first time and also equally likely to get any of the N balls from the bowl the second time, there are $N \times N$ equally likely combinations in all. How many of these are cases in which the first ball is black and the second is white? We know that any of the B black balls could be drawn the first time. For each of these balls, any of the W white balls could be drawn the second time. So, there are $B \times W$ ways in which we might draw a black ball first and a white ball second.

$$P((x_1 \text{ is black})\&(x_2 \text{ is white})) = \frac{BW}{NN}.$$

Notice that

$$\frac{BW}{NN} = \left(\frac{B}{N}\right)\left(\frac{W}{N}\right) = P(x_1 \text{ is black}) \times P(x_2 \text{ is white}).$$

In general, *the probability of several independent events is the product of their individual probabilities*. (Events are "independent" if the occurrence of one of them does not affect the probability of the others. In this case, the probability of getting a white ball on the second draw is not affected by whether we got a black ball on the first draw, since we are drawing with replacement.)

What is the probability of getting exactly two black balls from the first five draws (with replacement)? There are $N \times N \times N \times N \times N = N^5$ possibilities. There are $B \times B \times W \times W \times W = B^2 W^3$ ways of getting two black balls followed by three white balls. There are the same number of ways to get any other sequence pattern with exactly two black balls and three white balls (e.g., white-black-white-white-black). There are $\frac{5 \times 4}{2 \times 1} = 10$ such sequence patterns. So, the probability of getting exactly two black balls from the first five draws (with replacement) is $\frac{10 B^2 W^3}{N^5}$.

We can see this as follows. If there are five possible positions for the two black balls, then there are five choices for the first black ball. And once we choose the position of this first black ball, then there are only four possible positions for the second black ball. Thus, there are $5 \times 4 = 20$ possible placements for black ball 1 and black ball 2. But, this counts the positions of the black balls twice. That is to say, the pattern white-black-white-white-black could be obtained with black ball 1 in position 2 and black ball 2 in position 5, or with black ball 1 in position 5 and black ball 2 in position 2. Either possibility would give the pattern white-black-white-white-black. Thus, to take care of the double counting, we divide the count by 2.

In general, the number of ways of selecting r items from a set of n items is

$$nCr = \frac{n!}{r!(n-r)!}. \tag{2.1}$$

In this formula, $n! = n(n-1)(n-2)\cdots(2)(1)$ and is read "n factorial." To see this, the reasoning is similar to that in the special case we discussed above. In particular, if we select r items from a set of n items, then there are n choices for the first item, $n-1$ choices for the second item, and so on, with $n-r+1$ choices for the r-th item. This gives

$$n(n-1)\cdots(n-r+2)(n-r+1) = \frac{n!}{(n-r)!}.$$

But as before, we need to take care of multiple counting. Now, with r items, instead of double counting, we have counted each arrangement $r!$ times and so need to divide the count by $r!$, which gives Equation (2.1).

2.3 CONDITIONAL PROBABILITY

Suppose that five balls are drawn with replacement. What is the probability that exactly two black balls are drawn, given that there is at least one black ball? We write this conditional probability as $P(A|B)$, where A is the event of drawing exactly two black balls and B is the event of drawing at least one black ball.

To compute this conditional probability, we compare the number of possibilities in which at least one black ball is drawn with the number of possibilities in which exactly two black balls are drawn.

The number of possibilities in which at least one black ball is drawn must be equal to the total number of possibilities minus the number of possibilities in which only white balls are drawn. Since there are W^5 ways to draw only white balls, there are $N^5 - W^5$ ways to get at least one black ball.

We already calculated that there are $10B^2W^3$ ways to draw exactly two black balls when five balls are drawn with replacement. So, the conditional probability of drawing exactly two black balls, given that at least one black ball is drawn out of five drawn with replacement is

$$\frac{10B^2W^3}{N^5 - W^5}.$$

Suppose the balls are black or white and also plastic or glass. So there are four categories, black and plastic, black and glass, white and plastic, and white and glass. Suppose we know the number of balls of each of these categories, $NUM(BP)$, $NUM(BG)$, $NUM(WP)$, and $NUM(WG)$. From that information we can figure out how many balls in all there are, N, how many are black, how many are white, how many are glass, and how many are plastic. A ball is randomly selected. Given all that information, what is the conditional probability $P(B|G)$ that a ball is black, given that it is glass? It will be the fraction of the glass balls that are black. In other words,

$$P(B|G) = \frac{NUM(BG)}{NUM(G)}.$$

If we divide the numerator and the denominator of that fraction by the total number of balls, N, we get

$$\frac{\dfrac{NUM(BG)}{N}}{\dfrac{NUM(G)}{N}},$$

which is the same as

$$\frac{P(B \& G)}{P(G)}.$$

As a general rule, as long as $P(B) > 0$, we have

$$P(A|B) = \frac{P(A \& B)}{P(B)}.$$

This last equation is sometimes taken to be the definition of conditional probability, at least for the case in which $P(B) > 0$.

2.4 DRAWING WITHOUT REPLACEMENT

Suppose that we do not put back the balls after they are drawn. Instead, we choose randomly from the balls remaining in the bowl. Then the probabilities change after each drawing.

$$P(x_2 \text{ is white}|x_1 \text{ is white}) = \frac{W - 1}{N - 1}$$

and

$$P(x_2 \text{ is white}|x_1 \text{ is black}) = \frac{W}{N-1}.$$

What is the probability of getting four white balls in a row, drawing without replacement? Assuming that there are at least four white balls (i.e., $W \geq 4$), we have

$$P((x_1 \text{ is white})\&(x_2 \text{ is white})\&(x_3 \text{ is white})\&(x_4 \text{ is white}))$$

$$= \left(\frac{W}{N}\right)\left(\frac{W-1}{N-1}\right)\left(\frac{W-2}{N-2}\right)\left(\frac{W-3}{N-3}\right).$$

2.5 A CLASSIC BIRTHDAY PROBLEM

What is the probability that in a room with 23 people chosen at random, at least two will have the same birthday? Assume that no one in the room is born on February 29 and that a person is equally likely to be born on all the other dates.

The answer is 1 minus the probability that no two of them have the same birthday.

Assume that the people in the room can be ordered in some way and consider the probability that any given person does not have the same birthday as any of the earlier people in the order.

There are 364 possible days for the second person's birthday if it is to be different from the first person's birthday, so the probability that the second does not have the same birthday as the first is 364/365. By the same reasoning, assuming that the first two do not have the same birthday, the probability that the third does not have the same birthday as either of them is 363/365. If none of the first $N - 1$ people have the same birthday, then the probability that the Nth does not have the same birthday as any previous person is $(366 - N)/365$.

So, the probability that no two of the 23 have the same birthday is

$$\frac{364 \times 363 \times \cdots \times 343}{365^{22}} \approx 0.49$$

(the symbol $\approx$ means "is approximately equal to"). Then, the probability that at least two have the same birthday ≈ 0.51.

2.6 RANDOM VARIABLES

In all the previous examples (the focus of this chapter), there are a finite number of outcomes of a random experiment. The outcome of a random experiment is called a *random variable*. Often a random variable is denoted by X, where the capital letter is used to distinguish a random variable from the lower case x, which is used to denote a specific realization of the random variable X.

Suppose there are k possible outcomes for a random variable X and denote them by $a_1, a_2, \ldots, a_k$. We assume that these outcomes are mutually exclusive and exhaustive. That is, in any trial of the experiment, one and only one of the a_i occurs.

Suppose that the outcomes $a_1, a_2, \ldots, a_k$ occur with probabilities $p_1, p_2, \ldots, p_k$, respectively. That is to say, the probability of outcome a_i is p_i. Each probability p_i satisfies

$$0 \le p_i \le 1,$$

and since the outcomes are mutually exclusive and exhaustive, we have

$$\sum_{i=1}^{k} p_i = 1.$$

So, the random variable $X = a_i$ with probability p_i. This is often written as $P(X = a_i) = p_i$ or sometimes as $P_X(a_i) = p_i$. P_X is called the distribution of the random variable X. If the random variable X is understood and there is no confusion about the distribution one is considering, then sometimes the distribution is denoted simply by P and the probability of a_i is denoted simply by $P(a_i)$.

For example, in the case of B black balls and W white balls in a bowl with $N = B + W$ total balls, we have two outcomes. The outcome a_1 might denote the case where a black ball is drawn and a_1 might denote the case where a white ball is drawn. The corresponding probabilities are $p_1 = B/N$ and $p_2 = W/N$. If we toss a fair coin, the two outcomes are $a_1 =$ head and $a_2 =$ tail with probabilities $p_1 = p_2 = 1/2$.

If we roll a fair die, there are six outcomes $a_1, \ldots, a_6$, where a_i denotes the outcome that we roll i, and each $p_i = 1/6$. Equivalently, we write $P(1) = 1/6$, $P(2) = 1/6$, and so on. If the die is weighted in such a way that we never roll an even number and the odd numbers are equally likely, then $p_1 = p_3 = p_5 = 1/3$ and $p_2 = p_4 = p_6 = 0$. Then $P(1) = 1/3$, $P(2) = 0$, and so on.

The case of a die where the outcomes are associated with real numbers is very common. Another case that is very common is that in which the outcomes are vectors of real numbers. Both of these cases will be used throughout this book.

2.7 EXPECTED VALUE

Consider the case where the random variable X takes on real values. We define the *average* or *mean* or *expected value* of X, denoted by $E[X]$, as the weighted average of the possible outcomes where the weights are the probabilities corresponding to the outcomes. In particular, if the possible outcomes are $a_1, \ldots, a_k$

with probabilities $p_1, \ldots, p_k$, respectively, then

$$E[X] = \sum_{i=1}^{k} p_i a_i.$$ (2.2)

For example, if X denotes the outcome of rolling a fair die, then the possible outcomes are $1, 2, 3, 4, 5, 6$ and each of these has a probability of $1/6$. So, we have

$$E[X] = \sum_{i=1}^{6} p_i a_i$$

$$= \sum_{i=1}^{6} \frac{1}{6} i$$

$$= \frac{1}{6} + \frac{2}{6} + \frac{3}{6} + \frac{4}{6} + \frac{5}{6} + \frac{6}{6}$$

$$= 3.5.$$

On the other hand, suppose the die is weighted, as we discussed before, in such a way that the odd numbers are equally likely but the even numbers have a probability of 0. If X denotes the outcome of rolling this die, then

$$E[X] = \frac{1}{3} + 0 + \frac{3}{3} + 0 + \frac{5}{3} + 0$$

$$= 3.$$

Note that the expected value of X need not be the most likely value of X. In the case of a fair die, the expected value is 3.5, but the outcome of a fair die is never equal to 3.5. Intuitively, the expected value of X is what we might expect to get if we rolled the die many times and took the average of all the observed outcomes. This can be quantified in various ways through results known as the laws of large numbers, which we describe briefly in Section 3.7.

2.8 VARIANCE

Consider the following two random variables:

$$X_1 = \begin{cases} 1 \text{ with probability } 1/3 \\ 3 \text{ with probability } 1/3 \\ 5 \text{ with probability } 1/3 \end{cases}$$

$$X_2 = \begin{cases} 2 \text{ with probability } 1/3 \\ 3 \text{ with probability } 1/3 \\ 4 \text{ with probability } 1/3 \end{cases}$$

X_1 is the same random variable we considered in the previous section and has a mean of 3. We can easily check that X_2 also has a mean of 3. But in some sense, X_2 is tighter or less spread out about its mean. An important attribute of a random variable called the *variance* quantifies this idea of deviation from the mean.

As before, suppose X is a random variable for which the outcome a_i occurs with probability p_i for $i = 1, \ldots, k$. Let $\mu = E[X]$ denote the mean of X.

If we observe a_i, then the deviation from the mean is $a_i - \mu$, and we observe this deviation with probability p_i. This is observing the random variable $X - \mu$. If we simply took the mean of these deviations (i.e., the mean of $X - \mu$), then we would get zero because the positive contributions (when X is larger than μ) would cancel out the negative contributions (when X is smaller than μ).

To avoid this cancelation, we might consider the absolute deviation, namely $|X - \mu|$, but another more common and analytically more useful alternative is to consider the squared deviation $(X - \mu)^2$. This is a random variable that takes on the value $(a_i - \mu)^2$ with probability p_i. The expected value of this random variable is called the *variance* of X and is often denoted by σ^2 (or σ_X^2 or $\sigma^2(X)$ if it is necessary to make the underlying random variable explicit). That is to say, the variance σ^2 is defined as

$$\sigma^2 = E[(X - \mu)]^2, \tag{2.3}$$

where $\mu = E[X]$ is the mean of X. By some simple manipulations and properties of expectation, it can be shown that this can also be written as

$$\sigma^2 = E[(X - \mu)]^2 = E[X^2] - \mu^2. \tag{2.4}$$

So, when X takes on the value a_i with probability p_i for $i = 1, \ldots, k$, the expression for the variance of X can be written in any of the following ways:

$$\sigma^2 = \sum_{i=1}^{k} p_i (a_i - \mu)^2$$

$$= \sum_{i=1}^{k} p_i a_i^2 - \mu^2$$

$$= \sum_{i=1}^{k} p_i a_i^2 - \left(\sum_{i=1}^{k} p_i a_i \right)^2.$$

Returning to the random variables X_1 and X_2 that we defined at the beginning of this section, recall that both have a mean of 3. Hence, the variance of X_1 and X_2, denoted by σ_1^2 and σ_2^2, respectively, are given by

$$\sigma_1^2 = 1/3(1 - 3)^2 + 1/3(3 - 3)^2 + 1/3(5 - 3)^2 = 8/3$$

and

$$\sigma_1^2 = 1/3(2-3)^2 + 1/3(3-3)^2 + 1/3(4-3)^2 = 2/3.$$

so, as expected, the variance of X_2 is smaller than X_1.

Finally, we mention that the *standard deviation* of a random variable is just the square root of the variance. So, if σ^2 is the variance of a random variable, then σ is the standard deviation.

2.9 SUMMARY

In this chapter we reviewed some basic principles of probability theory. The probability of an event can range between 0 and 1. The probability of an event that is certain is 1 and the probability of an event that is impossible is 0. The probability of two mutually exclusive events is the sum of their probabilities. The probability that an event will not occur is 1 minus the probability that it will occur. The probability that two independent events occur is the product of the probabilities that each individual event occurs.

We discussed the notion of conditional probability and considered a number of cases, including the classic birthday problem. We then defined the notion of a random variable. We discussed the expected value and the variance of of a random variable.

2.10 APPENDIX: INTERPRETATIONS OF PROBABILITY

Albert shows you a possibly biased coin to be used for coin tossing. It can come up either heads or tails. What is the probability that it will come up heads on its next toss?

We can distinguish at least two or three (or more) ways of understanding this question, depending on whether it concerns subjective, objective, or epistemic probability.

Subjective probability has to do with your *degree of belief* in the proposition that the coin will come up heads, perhaps as measured by what bets you would accept or reject. If you could see no more reason to bet that the coin will come up heads than to bet that it will come up tails, then your degree of belief that the coin will come up heads would be 0.5 and that would be your subjective probability that the coin will come up heads.

On the other hand, although you would not know this, the coin might have been weighted so as to come up heads 60 percent of the time. In that case, the *objective probability* would be 0.6 that the coin will come up heads. The objective probability of a coin's coming up heads is (roughly) the limit of the frequency with which that would occur.

Epistemic probability, if there is such a thing, has to do with *rational* degree of belief, perhaps as measured by what bets it would be rational for you to accept or reject if you were to fully appreciate your evidence and were to reason absolutely correctly.

In this book, we are concerned always with objective probability, which is also called objective chance.

Objective probability is not the same as either subjective probability or epistemic probability. You may have no idea what the actual objective probability of an event is and not enough evidence to figure it out.

We said that objective probability is (roughly) the limit of the long-run frequency. But actually, objective probability is a primitive notion that cannot be defined in more primitive terms. Consider a fair coin that is tossed again and again forever, with no loss of metal, etc. In this infinite series of tosses, any infinite series of heads and tails is possible. The objective probability of any one of them is 0. With a fair coin, an infinite string of all heads is possible, but with a probability of 0.

Strictly speaking, then, we cannot identify the objective probability of getting heads with the long-run frequency of heads. But we can say that the objective probability approaches 1 that the long-run frequency of heads will approach the objective probability of heads.

The objective probability of an event is always a relative matter. It is the probability of that event's occurring in a certain idealized chance setup, for example, a certain person's tossing a coin on a particular occasion and its landing either heads up or tails up. At one level of abstraction, the probability of heads may be 1/2. At another level, given the particular way the coin is tossed, the existing air currents and possible landing surfaces, it may be causally determined how the coin lands, so the probability of heads may be 1 or 0. At the quantum level, it may be probabilistic again even with those details.

If the coin is tossed by a skilled magician who aims at the coin's coming up heads, the probability that the coin comes up heads may be quite high. If the coin is tossed by an average person, the probability may be 1/2.

To repeat, in this book we address only objective probabilities so conceived and we often make certain assumptions about these probabilities which imply that observed frequencies can provide evidence about objective probabilities.

2.11 QUESTIONS

1. Given an opaque bowl containing six all blue balls, seven all red balls, and nothing else, if a ball is withdrawn at random, what is the probability that the ball is blue?

2. True or False: For any event A, we have $P(\text{not } A) = 1 - P(A)$?

3. If the events A and B are independent, what is $P(A\&B)$?

4. True or False: For any events A and B, we have $P(A \text{ or } B) = P(A) + P(B)$.

5. True or False: For any events A and B, $P(A \text{ and } B) < P(A)$

6. Why is it true that the number of ways of selecting r items from a set of n items is $nCr = \frac{n!}{r!(n-r)!}$?

7. Suppose that the probability of drawing a white ball from a bowl containing only black and white balls is $w/(w + b)$, where w is the number of white balls and b is the number of black balls. What is the probability of drawing a black ball? What is the probability of drawing either a white ball or a black ball? What is the probability of drawing a ball that is neither a white ball nor a black ball? Show that the probability of drawing a white ball is equal to or greater than zero and less than or equal to one.

8. Given the same suppositions as in the previous question, suppose also that the balls come in two shapes, round and egg-shaped, where the number of round balls is r and the number of egg-shaped balls is e (where $w + b = r + e$). What is the probability that a randomly drawn ball will be either white or egg-shaped? Explain.

9. What is the difference between drawing with replacement and drawing without replacement? Does this difference affect the probabilities in any way? Explain.

10. Using the suppositions in question 3, suppose also that a ball is randomly drawn and then replaced, then again a ball is randomly drawn and replaced. What is the probability that the same ball is drawn both times? What is the probability that the two balls have the same features (color and shape)? What is the probability that the two balls share one but only one feature?

11. True or False: If we toss an unbiased coin (50% heads and 50% tails) 100 times, then by the Law of Large Numbers we will have exactly 50 heads and 50 tails.

12. True or False: if a fair coin is tossed infinitely many times, each time with a probability of 0.5 for getting heads and 0.5 for getting tails, the probability that heads will come up every time HHHH... is less than the probability that heads and tails will alternate for ever HTHTHT....

13. True or False: Depending on the circumstances, the subjective probability of an event can be less than, equal to, or greater than its statistical probability.

2.12 REFERENCES

The word "probability" is used to describe several distinct, though related, ideas such as degree of belief and chance. Much of the early interest in formalizing probability arose largely from a desire to understand so-called "games of chance," where the outcome of the game was determined to some extent by events deemed to be random, such as the roll of a die or the flip of a coin. Perhaps the earliest work in this regard was that of Cardano in the 1500s. A more rigorous and formal treatment was initiated through correspondences between the French mathematicians Pascal and Fermat in the 1600s. Their work is generally considered the beginning of the mathematical theory of probability, and was the prevailing view of probability among mathematicians until the twentieth century, with contributions from such luminaries as Huygens, Bernoulli, de Moivre, Laplace, Gauss, and others.

In the 1930s, the relative frequency view of probability emerged, largely through the work of Von Mises and Fisher. At about the same time as the relative frequency view was emerging, an axiomatic approach was being developed that gave shape to the widely accepted form of the current probability theory. This axiomatic approach was due in large part to the efforts of Kolmogorov. The classical view of probability is subsumed as a very special case, and the relative frequency approach is reconciled via results we briefly mentioned known as the "Laws of Large Numbers."

There are many good books on probability theory. Everitt (1999) is a nice "guide." Feller (1968) is a classic. Hacking (1965) is a philosophically useful guide. Hacking (1984) is a historical account. Rabinowitz (2004); Bertsekas and Tsitsiklis (2008), and Tijms (2007), and Ross (2009) are some fairly recent books at different levels.

Bertsekas D, Tsitsiklis J. Introduction to probability. 2nd ed. Belmont, MA: Athena Scientific; 2008.

Everitt BS. Chance rules: an informal guide to probability, risk, and statistics. New York: Copernicus, Springer-Verlag; 1999.

Feller W. Volume 1, An introduction to probability theory and its applications. 3rd ed. New York: Wiley; 1968.

Hacking I. The logic of statistical inference. Cambridge: Cambridge University Press; 1965.

Hacking I. The emergence of probability. Cambridge: Cambridge University Press; 1984.

Rabinowitz L. Elementary probability with applications. Wellesley, MA: AK Peters; 2004.

Ross SM. A first course in probability. 8th ed. Upper Saddle River, NJ: Prentice-Hall; 2009.

Tijms H. Understanding probability: chance rules in everyday life. Cambridge: Cambridge University Press; 2007.

CHAPTER 3

Probability Densities

In Chapter 2, we explained some of the elementary mathematics of probability in the discrete setting, in which the number of outcomes of an experiment is finite. In this chapter, we discuss some of the basics of probability in the case where the outcomes take values in a continuum.

3.1 AN EXAMPLE IN TWO DIMENSIONS

Alice shoots at a target. Let us suppose that her bullet can hit any point on the target (imagine that her bullet is infinitesimally small). Perhaps the probability that she will hit a point within 1 inch of the center of the target is 0.5 and the probability that she will hit a point within 2 inches of the center is 0.9.

The probability that Alice will hit the exact center of the target is 0, and it is so also for any other point on the target. But there is a sense in which she is more likely to hit the center point than some particular point further out. We can capture this sense in terms of a probability density over the plane $\mathbf{R}^2$ (Figure. 3.1).

To explain this idea of probability density precisely, we start with the 1-dimensional case and then return to higher dimensions in Section 3.4.

3.2 RANDOM NUMBERS IN [0,1]

What does it mean to select a random number from the interval [0,1] with all the numbers being equally likely? What is the probability of selecting a particular number, say 0.5?

If all the numbers are equally likely, then the probability of selecting 0.5 is the same as the probability of selecting 0.75 or $\pi - 3$. But since each of these outcomes is mutually exclusive, then the probability of each must be 0. Otherwise, since there are infinitely many points in the interval [0,1], we would get a contradiction. That

An Elementary Introduction to Statistical Learning Theory, First Edition.
Sanjeev Kulkarni and Gilbert Harman.
© 2011 John Wiley & Sons, Inc. Published 2011 by John Wiley & Sons, Inc.

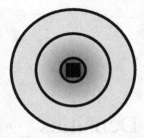

Figure 3.1 Probability density of hitting a point on a target.

is, suppose that the probability of selecting each point is some value $p > 0$. Then, if a_1 and a_2 are two distinct points in [0,1], then $P(a_1 \text{ or } a_2) = 2p$. Likewise, if $a_1, \ldots, a_N$ are N distinct points in [0,1], then $P(\cup_{i=1}^{N} a_i) = Np$. But for a large enough N, we have $Np > 1$, while probabilities are required to be between 0 and 1.

So, the probability of selecting any particular number is 0 yet the probability of selecting *some* number in [0,1] is 1.

Now consider the case where with a probability of 1/3, we get a random number between 0 and 1/2 with all the numbers in the interval [0,1/2] being equally likely. With a probability of 2/3, we get a number between 1/2 and 1, with all the numbers in the interval (1/2,1] also being equally likely.

Using an argument similar to above, the probability of observing any particular number is 0. So in what sense are the numbers in the interval (1/2,1] twice as likely as those in the interval [0,1/2]? To answer this question, it makes more sense to consider the probability of sets of points rather than the probability of seeing an individual point. One way to do this is to use density functions, which we describe in the next section.

3.3 DENSITY FUNCTIONS

A very useful way to work with probabilities in the continuous case is to use *density functions*. A function $p(x)$ that is non-negative and with area under $p(x)$ equal to 1 is a valid density function. Strictly speaking, we need $p(x)$ to satisfy another rather technical condition called *measurability*. We briefly discuss the notion of measurable sets and measurable functions in the appendix, but a detailed discussion is beyond the scope of this book. We assume that all the functions and sets we deal with are measurable.

Given a density function $p(x)$, the probability of an interval A is given by the area under the function $p(x)$ over the interval A. This applies to more general sets as well. That is to say, for a general (measurable) set A, the probability of A, denoted $P(A)$, is just the area under $p(x)$ over A. In calculus, this is called the

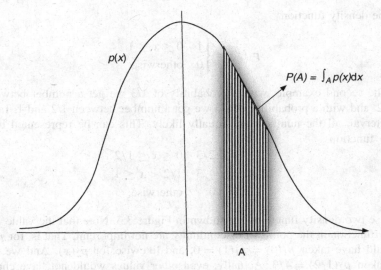

$$P(A) = \int_A p(x)\,dx$$

Figure 3.2 Density and probability.

integral of $p(x)$ over A and is written as

$$P(A) = \int_A p(x)\,dx. \tag{3.1}$$

This is depicted in Figure 3.2.

With this notation, the two conditions required for a (measurable) function $p(x)$ to be a valid density function are

$$p(x) \geq 0 \quad \text{for all } x \tag{3.2}$$

and

$$\int_{\mathbf{R}} p(x)\,dx = 1. \tag{3.3}$$

The first condition guarantees that $P(A) \geq 0$ for any set A. The second condition guarantees that the probability of the whole real line (i.e., the probability of selecting *some* real number) is 1. The two conditions together ensure that for any set A, we have $0 \leq P(A) \leq 1$ as required for the probability of a set.

Returning to the examples discussed in Section 3.2, consider the case where a number is selected from the interval [0,1] with all the numbers being equally likely. What this means is that given any small interval contained in [0,1], the probability of the interval is the same regardless of the exact position. This can be represented with the density function

$$p_1(x) = \begin{cases} 1 & 0 \le x \le 1 \\ 0 & \text{otherwise.} \end{cases}$$

In the second example, with a probability of 1/3 we get a number between 0 and 1/2, and with a probability of 2/3 we get a number between 1/2 and 1. In each half interval, all the numbers are equally likely. This can be represented by the density function

$$p_2(x) = \begin{cases} 2/3 & 0 \le x \le 1/2 \\ 4/3 & 1/2 < x \le 1 \\ 0 & \text{otherwise.} \end{cases}$$

These two density functions are shown in Figure 3.3. Note that the value of the density function at the points of discontinuity are not important. That is, for $p_1(x)$, we could have taken $p_1(0) = p_1(1) = 0$, and likewise for $p_2(x)$. And we could have taken $p_2(1/2) = 4/3$. Actually, even other values would not have changed things. The value of a density at any finite number of points is not important, since this does not change the integral (area under $p(x)$) over any set A, and so does not change the probability $p(A)$.

The value $p(x)$ does not represent the probability of the point x. As we have already mentioned, in the continuous case the probability of any individual point is generally zero, while for many points we may have $p(x) > 0$. Also, $p(x)$ can take on values greater than 1.

For an interpretation of $p(x)$, roughly we can think of the "probability mass" as being spread out over the real line with $p(x)$ representing how concentrated the mass is at x. More precisely, let $B(x, \epsilon)$ denote the interval $(x - \epsilon, x + \epsilon)$. An approximation to the density at x is given by $P(B(x, \epsilon))/(2\epsilon)$. As we let ϵ get smaller, we get better approximations to $p(x)$. In this limit, we get

$$p(x) = \lim_{\epsilon \to 0} \frac{P(B(x, \epsilon))}{2\epsilon}. \tag{3.4}$$

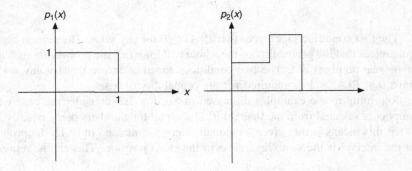

Figure 3.3 More density and probability.

3.4 PROBABILITY DENSITIES IN HIGHER DIMENSIONS

Returning to the 2-dimensional case of Alice shooting at a target, let $p(\overline{x})$ denote a probability density function that captures where on the target Alice's shot is likely to fall. As in 1-dimension, $p(\overline{x})$ needs to satisfy the conditions

$$p(\overline{x}) \geq 0 \quad \text{for all } \overline{x} \tag{3.5}$$

and

$$\int_{\mathbf{R}^2} p(\overline{x})\, d\overline{x} = 1. \tag{3.6}$$

Given a density function $p(\overline{x})$, the probability $P(S)$ that Alice's shot falls in a certain area S on the target is given by

$$P(S) = \int_S p(\overline{x})\, d\overline{x}. \tag{3.7}$$

The interpretation of $p(\overline{x})$ as a density is also similar to the 1-dimensional case. That is to say, fix a point $\overline{x}$ on the target and let $B(\overline{x}, \epsilon)$ denote the ball of radius ϵ with $\overline{x}$ at its center. We calculate the probability $P(B(\overline{x}, \epsilon))$ that Alice's shot will hit somewhere in the ball, and then divide that probability by the area of the ball, which is given by $\text{area}(B(\overline{x}, \epsilon)) = \pi\epsilon^2$. Then, as in the one-dimensional case, the probability density $p(\overline{x})$ at the point $\overline{x}$ equals

$$\lim_{\epsilon \to 0} \frac{P(B(\overline{x}, \epsilon))}{\text{area}(B(\overline{x}, \epsilon))}. \tag{3.8}$$

Likewise, we can consider probability density functions in $\mathbf{R}^d$ as well. The development is completely analogous to the 2-dimensional case. In particular, Equations (3.5), (3.6), (3.7), and (3.8) hold with the only change being that $\mathbf{R}^2$ is replaced with $\mathbf{R}^d$, and that the area of the ball $B(\overline{x}, \epsilon)$ is replaced with the volume of the ball.

3.5 JOINT AND CONDITIONAL DENSITIES

If we are interested in two random variables x and y, then a joint probability density function $p(x,y)$ describes the joint behavior of the two variables. To be a valid joint density function, $p(x,y)$ needs to satisfy the conditions

$$p(x, y) \geq 0 \quad \text{for all } x, y \tag{3.9}$$

and

$$\iint p(x, y)\, dx\, dy = 1. \tag{3.10}$$

as we would expect.

As with probabilities, we can consider conditional densities. The conditional density of x, given y, denoted $p(x|y)$, is given by

$$p(x|y) = \frac{p(x, y)}{p(y)} \tag{3.11}$$

whenever $p(y) > 0$.

3.6 EXPECTED VALUE AND VARIANCE

As in the discrete case, the mean (or average) of a random variable in the continuous case is an important quantity. The term *expected value* is used in both cases. Also, the expected value can be defined for a random variable that takes values in a higher dimensional space.

If $p(\overline{x})$ is a probability density function for a random variable $\overline{X}$ that takes values in $\mathbf{R}^d$, then the expected value of $\overline{X}$ is defined as

$$E[\overline{X}] = \int_{\mathbf{R}^d} \overline{x}\, p(\overline{x})\, \mathrm{d}\overline{x}. \tag{3.12}$$

Equation (3.12) is analogous to the computation in the discrete case. In the discrete case, as we see from Equation 2.2, the expected value is computed by multiplying the possible values that can arise by the corresponding probability and adding these up. Equation (3.12) reflects the natural extension to the continuous case in which the probability is replaced by the probability density function and the sum is replaced by the integral. Of course, if $\overline{X}$ takes on values in $\mathbf{R}^d$ then $E[\overline{X}]$ is also an element of $\mathbf{R}^d$.

Likewise, the notion of the variance of $\overline{X}$ can also be naturally extended to the continuous case. But the extension to higher dimensions (for both the discrete and continuous cases) is a little more involved. In the scalar case for which $d = 1$ so that X takes on values in $\mathbf{R}$, the variance, denoted by σ^2, is again defined as

$$\sigma^2 = E[(X - \mu)^2] = E[X^2] - \mu^2.$$

But now, the expectations are written in terms of integrals involving $p(x)$, that is,

$$\sigma^2 = \int_{\mathbf{R}} (x - \mu)^2 p(x)\, \mathrm{d}x = \int_{\mathbf{R}} x^2 p(x)\, \mathrm{d}x - \mu^2.$$

In higher dimensions, the extension of the variance is actually a $d \times d$ matrix called the *covariance matrix*. The reason we need this is that for $\overline{x} \in \mathbf{R}^d$, it is not clear how to take the square of $\overline{x}$. If the vector $\overline{x}$ is written in terms of its components as $\overline{x} = (x_1, \ldots, x_d)$, then the density $p(\overline{x})$ can be thought of as a joint density in terms of the scalars $x_1, \ldots, x_d$. If we think of $\overline{x}$ as a column vector, and let $\overline{x}^T$ denote the transpose of $\overline{x}$, namely $\overline{x}$ as a row vector, then the product

$\overline{x}\,\overline{x}^T$ is a $d \times d$ matrix whose (i, j) entry is $x_i x_j$. Likewise, if $\overline{\mu} = E[\overline{X}]$ denotes the expected value of $\overline{X}$, then $\overline{\mu} \in \mathbf{R}^d$ and $\overline{\mu}\,\overline{\mu}^T$ is a $d \times d$ matrix. The *covariance matrix* of $\overline{X}$, denoted Σ, is defined as

$$\Sigma = E[(\overline{X} - \overline{\mu})(\overline{X} - \overline{\mu}^T] = E[\overline{X}\,\overline{X}^T] - \overline{\mu}\,\overline{\mu}^T.$$

The diagonal elements of Σ correspond to the variance of the respective components of $\overline{X}$, that is, the (i, i) entry of Σ is equal to the variance of the i-th component of $\overline{X}$.

3.7 LAWS OF LARGE NUMBERS

Consider the experiment of drawing black and white balls from a bowl that we discussed in Chapter 2. Suppose we do not know the probability of choosing a black ball (i.e., the fraction of black balls), and we wish to use data to estimate this probability.

We might try drawing many balls with replacement and estimate the unknown probability by the fraction of black balls that we observe among all the balls we draw. As we draw more and more balls, the probability that the observed frequency of black balls is *exactly* the same as the fraction of balls in the bowl that are black goes down. But the probability that the observed frequency is approximately the same as that fraction goes up in the following sense.

Consider a hypothesis H that the observed frequency of black balls drawn from the bowl differs from the actual proportion of black balls in the bowl by less than some small number, ϵ. As we draw more and more balls, with replacement, the probability of any such hypothesis H increases and approaches 1 in the limit.

More precisely, we can define a random variable X so that $X = 1$ if an event of interest occurs and $X = 0$ otherwise. Suppose p is the probability of the event of interest. For example, the event might be "we observe a black ball," in which case p denotes the probability of observing a black ball. Suppose we run the experiment many times, and denote the random variable by X_1 the first time we run the experiment and by X_2 the second time we run the experiment, and so on. We assume that the random variables $X_1, X_2, \ldots$ are independent.

If we run the experiment n times, then the number of times the event occurs is $X_1 + X_2 + \cdots + X_n$. Let $\hat{p}_n$ denote the estimate using these n observations of the probability p that the event occurs. That is,

$$\hat{p} = \frac{X_1 + X_2 + \cdots + X_n}{n}.$$

It can be shown that as we get more and more observations, the estimate $\hat{p}_n$ converges to p in various ways. For example, given any $\epsilon > 0$, it can be shown that

$$\text{Prob}\{|\hat{p}_n - p| > \epsilon\} \to 0 \text{ as } n \to \infty.$$

This is a special case of estimating the expected value of a random variable by averaging a number of independent observations. Specifically, suppose X is a real-valued random variable and let $\mu = E[X]$ denote the expected value of X. For example, X might be the outcome of rolling a weighted die, and we might try to estimate μ by simply averaging a number of independent observations. That is, we roll the die many times and take the average of the outcomes as an estimate for μ. That is to say, our estimate $\hat{\mu}_n$ after n observations is given by

$$\hat{\mu}_n = \frac{X_1 + X_2 + \cdots + X_n}{n}.$$

It can be shown that for any $\epsilon > 0$ we have

$$\text{Prob}\{|\hat{\mu}_n - \mu| > \epsilon\} \to 0 \text{ as } n \to \infty.$$

These results are known as the Weak Law of Large Numbers, and a stronger statement can be made called the Strong Law of Large Numbers. Furthermore, it is possible to get results bounding various probabilities or on rates of convergence. Details are available in many probability and statistics textbooks (e.g., Feller, 1968, pp. 228–247).

3.8 SUMMARY

In this chapter, we reviewed some basic tools of probability in the continuous case. A probability density function captures the likelihood of occurrence of different values of the underlying random variable. A density function $p(\overline{x})$ must be non-negative and have the total area under the function equal to 1. The probability that the outcome belongs to a set A is given by area under the $p(\overline{x})$ over the set A. The value $p(\overline{x})$ can be obtained by dividing the probability of a small ball centered at $\overline{x}$ by the volume of the ball and taking the limit as the radius of the ball tends to zero. We discussed the notion of joint densities and conditional densities when two or more random variables are involved. The notion of expected value of a random variable extends naturally to the continuous case where a sum is replaced with an integral involving the probability density. The notion of variance extends naturally in the 1-dimensional case (i.e., a scalar-valued random variable), but in the higher dimensional case, the extension of variance gives rise to the covariance matrix. Finally, we ended with a brief discussion of laws of large numbers which hold for both continuous and discrete random variables.

3.9 APPENDIX: MEASURABILITY

A technical but fundamentally very important notion that we mentioned but did not discuss is measurability. A detailed discussion is beyond the scope of this book, but in this appendix, we briefly discuss this idea.

Consider subsets of the real line. Given a particular subset A, a natural question is how large is A? If A is an interval, a natural measure of the size of A is its length. So, if A is the interval $[0,1]$, its size is 1, while if A is the interval $[-1,2.5]$, then the size of A is 3.5. Note that the size of a single point is 0, so that it does not matter whether we consider open or closed intervals in the examples above. That is, the intervals $(-1,2.5)$, $(-1,2.5]$, and $[-1,2.5)$ also have size 3.5. If A consists of a finite union of intervals, then the size of A is just the sum of the sizes of the intervals comprising A.

But what about more complicated subsets? For example, suppose we restrict ourselves to subsets of $[0,1]$. What is the size of the rational numbers in $[0,1]$? What about the irrational numbers in $[0,1]$? Making this notion of size precise for general subsets is more difficult and subtle than it might first appear, and measure theory provides one approach.

It turns out that if we insist on some natural properties that our notion of size should satisfy, then there is no consistent way to assign a size to *every* subset. However, if we restrict attention to a reasonable collection of subsets, then we can assign a size (or measure) to each subset in a consistent way that satisfies the natural properties. This collection of subsets for which we can assign a size are called the measurable sets.

The collection of sets S to which a measure will be assigned is required to be what is called a σ-algebra. S is said to be a σ-algebra if S is nonempty and is closed under complements and countable unions. These last two conditions mean that if $A \in S$ then $A^c \in S$, and if $A_i \in S$ for $i = 1, 2, \ldots$ then $\cup_{i=1}^{\infty} A_i \in S$. ($A^c$ is the complement of A, the set of points not in A.)

Once we have a σ-algebra S, a measure μ is a function that assigns a real number or infinity to each member of S that satisfies the following properties: non-negativity, zero measure for the empty set, and countable additivity. Mathematically, these conditions are, respectively: $\mu(A) \geq 0$ for every $A \in S$; $\mu(\phi) = 0$, where ϕ is the empty set; and if $A_i \in S$ for $i = 1, 2, \ldots$, then $\mu(\cup_{i=1}^{\infty} A_i) = \sum_{i=1}^{\infty} \mu(A_i)$.

At first it is not clear why we cannot assign a size (measure) in a meaningful way to every set. For example, suppose we think about Alice throwing (infinitesimally small) darts at a dart board in the shape of the unit square with each point equally likely. Given any subset A of the unit square, Alice's dart either lands in A or not. The measure of A is just the probability that the dart lands in A.

However, a rather forceful counterargument came in 1905 when Vitali constructed a nonmeasurable set. That is, he constructed a set that under seemingly natural assumptions required for a measure of the size of sets together with the Axiom of Choice leads to a contradiction. This suggests the need to restrict the measure to a suitable collection of subsets (e.g., a σ-algebra S). Another very striking result that illustrates the complexities of defining measure is the Banach–Tarski paradox. In 1924, they showed that a ball in 3 dimensions can be decomposed into a finite number of pieces that can be moved, rotated, and fitted back together to produce two balls identical to the original! This result is sometimes informally described by saying "a pea can be chopped up and reassembled into the sun."

Resolving these types of problems requires us to abandon some of the properties we naturally expect of a measure, alter the axioms of Zermelo–Fraenkel set theory, or accept that there are nonmeasurable sets. The standard approach is to take the last option.

3.10 QUESTIONS

1. True or false: A probability density can take on values greater than 1.

2. True or false: A probability density can take on values less than 0.

3. Given the probability density function $f(x)$, what is the expression for $P(a \le x \le b)$ (that is, the probability that x takes a value in the range $[a,b]$)?

4. If $p_1(x)$ and $p_2(x)$ are two probability density functions, consider the function $p_\lambda(x) = \lambda p_1(x) + (1 - \lambda)p_2(x)$.

    (a) Is it true that $p_\lambda(x)$ is a valid density function for any $\lambda \in [0, 1]$? Why or why not?

    (b) What about for $\lambda \notin [0, 1]$? Why or why not?

5. Consider the function defined by

$$p(x) = \begin{cases} 2 - 2x & 0 \le x < 1 \\ 0 & \text{otherwise.} \end{cases}$$

    (a) Sketch $p(x)$.
    (b) Is $p(x)$ a valid probability density function? Why or why not?
    (c) What is the probability that $x \ge 1/2$?
    (d) What is the probability that $x = 1/2$?
    (e) For very small ϵ, what is the probability that x is within ϵ of $1/2$?

6. If the probability that Alice hits any given point on the target is the same as the probability that she hits any other given point, does that imply that she is as likely to hit a point within 1 inch of the center as she is to hit a point within a 1-inch circle further away from the center? Explain.

3.11 REFERENCES

As mentioned in the references to Chapter 2, the axiomatic approach to probability was due in large part to the efforts of Kolmogorov. This approach is based on the measure theory, which was developed by Borel, Lebesgue, and other mathematicians in the late 1800s and in the early 1900s. A probability measure is a special case of a general measure in which the measure of the whole set is one.

In addition to the books on probability mentioned in the references to Chapter 2, some more advanced books that provide a very rigorous treatment of probability based on measure theory include Billingsly (1995); Chung (2000); Dudley (2002).

Billingsly P. Probability and measure. 3rd ed. New York: Wiley Interscience; 1995.

Chung KL. A course in probability theory. 2nd ed. Amsterdam: Academic Press; 2000.

Dudley RM. Real analysis and probability. 2nd ed. Cambridge: Cambridge University Press; 2002.

Feller W. An Introduction to probability theory and its applications. 3rd ed. New York; Wiley; 1968.

CHAPTER 4

The Pattern Recognition Problem

Armed with some tools from probability that we discussed in the previous chapters, we can now begin discussing a statistical formulation and related results for learning and pattern recognition. These methods are widely used and have been extremely successful in many applications.

The problem we describe in this chapter is the two-class pattern recognition problem. This problem contains most of the essential features useful in a wide range of applications and for a rich and useful theory. This formulation is used throughout the book, except in Chapters 14 and 15, where we consider the estimation problem.

4.1 A SIMPLE EXAMPLE

Consider the following simple example of a pattern recognition problem. Suppose we have parts going along an assembly line and we wish to design an automatic inspection system. Some of the parts may be defective, while others are properly manufactured. Instead of having a person watch and inspect the items as they go by, we prefer a system that will automatically classify the items as either "good" (properly manufactured) or "bad" (defective). We use the label "1" for good objects and "0" for bad objects.

If we have no other information and no criteria, there is not much we can do besides random guessing. The approach taken in statistical pattern recognition is to assume that we first observe some features of each object that we wish to classify, perhaps color, length, width, color, and weight. We then use the observed values of the features to help in classifying the object. Moreover, it is assumed that we have a probabilistic model for the objects and the relationship between the measured features and the object class.

For example, in the inspection problem, suppose we know that bad items/ parts occur with some probability $P(0)$ and good parts occur with probability

An Elementary Introduction to Statistical Learning Theory, First Edition.
Sanjeev Kulkarni and Gilbert Harman.
© 2011 John Wiley & Sons, Inc. Published 2011 by John Wiley & Sons, Inc.

$P(1) = 1 - P(0)$. Also, suppose we measure a feature that can take on one of six values $v_1, v_2, \ldots, v_6$. The probabilistic model tells us not only how often good and bad parts occur but also the probability of observing each feature value for both good and bad parts. These probabilities are denoted by $P(v|0)$ and $P(v|1)$, and are conditional distributions for the feature value, given the class of the object.

We should expect that certain feature values are more likely if the part is bad, and other feature values are more likely if the part is good. We certainly expect that the distributions $P(v|0)$ and $P(v|1)$ are not the same. Otherwise, observing the feature would not be very helpful in trying to classify the object as good or bad.

Some immediate questions that arise are "How do we design good classification (or decision) rules?," "How should we measure the performance of a rule," and "How do we even formalize the notion of 'decision rule'?"

4.2 DECISION RULES

Once we observe the values of an object's features, we make a guess as to the class to which the object belongs. For example, in the inspection problem, if we observe the value v_1 for the feature, we may guess that the item is defective (belongs to class 0), and if we observe any other value, then we might guess that the object is good (belongs to class 1). This is one possible *decision rule* (or classification rule, or classifier).

Many other decision rules are possible. A very simple rule is to guess that the object always belongs to class 1 (i.e., is good). Another rule might be to guess that the object belongs to class 1 if the observed feature value is larger than some fixed number and that it belongs to class 0 otherwise.

In general, we observe several features of an object, say d features in all, and our observation is represented by a feature vector $\overline{x} = (x_1, x_2, \ldots, x_d)$. Here x_1 represents the measured value of the first feature, x_2 represents the measured value of the second feature, and so on up to x_d. A decision rule is really just a map that assigns either 0 or 1 to each possible feature vector.

If Ω denotes the set of all possible feature vectors, then a decision rule c maps Ω to $\{0, 1\}$, that is, $c : \Omega \to \{0, 1\}$. Also, any such map corresponds to a decision rule, where $c(\overline{x})$ indicates the decision when feature vector $\overline{x}$ is observed.

Equivalently, we can think of a decision rule as a partition of Ω into two sets Ω_0 and Ω_1. Upon observing any feature vector in Ω, we must decide either 0 or 1. Ω_0 consists of precisely those feature vectors for which we decide 0, and likewise Ω_1 consists of those feature vectors for which we decide 1. That is, we decide 0 if $\overline{x} \in \Omega_0$ and we decide 1 if $\overline{x} \in \Omega_1$.

In most of this book, we consider the case where features take on real values and so the feature vector $\overline{x} = (x_1, x_2, \ldots, x_d)$ is an element of $\mathbf{R}^d$, where d is the number of features. In many cases, only certain values might be allowed so that there are only finitely many possible feature vectors $\overline{x}$. In other cases, we might think of each of the features x_i as being allowed to take on any real value. In this case, $\Omega = \mathbf{R}^d$, and a decision rule is a partition of this d-dimensional Euclidean space. See Figure 4.1 for a partition in $\mathbf{R}^2$.

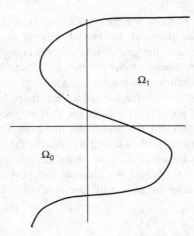

Figure 4.1   A decision partition in $\mathbf{R}^2$.

But, in other cases, we might have only finitely many possible feature values. Or we might have categorical features that are not naturally thought of as real values. In this case, Ω might be something other than $\mathbf{R}^d$, and might even be a finite set.

Another way to think about a decision rule is as a subset of Ω. If we know all the feature vectors for which we decide 1, then the rest must be those for which we decide 0 (since there are just two classes and we are assuming that we make a decision for all observations). So indicating that the portion of Ω that gets mapped to 1 completely defines a decision rule. Therefore, in the two-class case, the set of all decision rules can also be identified with the set of all subsets of Ω. This is called the "power set" of Ω.

The power set of Ω (and hence the set of all possible decision rules) is typically extremely rich. Even if Ω is finite with modest cardinality, the power set is huge. For example, if there are only 100 possible feature vectors, then there are 2^{100} (which is more than 10^{30}) possible decision rules! In some applications, there may be thousands or millions of possible feature vectors, and as we said before, in many cases we think of the features as taking on a continuous range of values so that the feature space is uncountably infinite.[1]

For the case where the features take on a continuous range of values, we often take $\Omega = \mathbf{R}^d$ or some appropriate subset of $\mathbf{R}^d$. In this case, for various technical/mathematical reasons, we do not consider the set of *all* possible subsets of $\mathbf{R}^d$ as potential decision rules. Instead, we restrict decision rules to be a special type of subset called a *measurable* set. This set is extremely rich and includes all sets normally found in practice, but the collection of measurable sets is not the power set of $\mathbf{R}^d$. In Section 3.9, we briefly discussed the notion of measurability. However, these issues are outside the scope of this book, and we assume that all sets and functions we deal with are measurable.

---

[1]The appendix to this chapter explains the meaning of "uncountably infinite."

4.3 SUCCESS CRITERION

Out of the enormous collection of all possible decision rules, we would like to select one that performs well. But first we need to specify what we mean by "performs well." That is, we need to specify how we measure the performance of a decision rule c.

First, notice that since we have assumed a probabilistic model for the objects and the connection between the observed features and the classification of the object, in general, no decision rule can always be right.

In light of the probabilistic model, one natural success criterion is the probability that we make a correct decision. We would like this probability to be as large as possible. Equivalently, we would like the probability of an incorrect classification (called the probability of error) to be as small as possible.

To proceed further, we would like an expression for the probability of error of a decision rule c. There are two ways in which the decision rule c can make an error: (i) the true class of the object is 0 but c classifies the object as 1 or (ii) the true class of the object is 1 but c classifies the object as 0.

Consider case (i). The probability that the object belongs to class 0 is just the prior probability $P(0)$. Given that the object belongs to class 0, we will make an error if and only if the observed feature vector belongs to Ω_1, since this will mean that our decision is 1. This conditional probability can be written as $P(\Omega_1|0)$. The probability that event (i) happens is then $P(0)P(\Omega_1|0)$.

Similarly, for case (ii), the probability that the object belongs to class 1 is $P(1)$, and given that the object belongs to class 1, we make an error if and only if the observed feature vector belongs to Ω_0. Then, the probability that event (ii) happens is $P(1)P(\Omega_0|1)$.

Since events (i) and (ii) are mutually exclusive, the probability of error of a rule c can be written as

$$\text{Probability of error of } c = P(0)P(\Omega_1|0) + P(1)P(\Omega_0|1).$$

4.4 THE BEST CLASSIFIER: BAYES DECISION RULE

We are now in a position to discuss the best decision rule. Recall the model so far. There are two classes 0 and 1 with prior probabilities $P(0)$ and $P(1)$. The observed feature values of the object are related to the class of the object through conditional probabilities $P(\overline{x}|0)$ and $P(\overline{x}|1)$. We observe a feature vector $\overline{x}$ and will decide either 0 or 1. We need a decision rule that minimizes the probability of error.

For now, we are assuming that we know $P(0)$, $P(1)$, $P(\overline{x}|0)$, and $P(\overline{x}|1)$ exactly. With this knowledge, the problem is really one of statistical decision theory rather than learning. That is, we are assuming everything about the model for the environment is known and we simply want to understand how to use this knowledge. This will give us a benchmark or target for what we might try to achieve if the distributions are not known and instead have to be learned from data.

First, consider a very simple case in which we do not observe any features of the object at all. In this case, the only information we have on which to base a decision are the prior probabilities $P(0)$ and $P(1)$. What class should we decide? If we want to minimize the probability of error, then no matter what object we see, our best guess is simply to always decide the class with higher prior probability. If $P(0) > P(1)$, then we should always guess that an object comes from class 0, while if $P(1) > P(0)$, then we should always guess that an object comes from class 1. With this decision rule, the probability that we make a correct decision is $\max(P(0), P(1))$ and the probability of error is $\min(P(0), P(1))$. Notice that if $P(0) = P(1) = 1/2$, the probability of error is 1/2, as is the probability of a correct decision. This is to be expected, since in this case we have no advantage over random guessing, so it does not matter what we decide.

Now, let us consider what to do if we observe a feature vector. If we were able to get conditional probabilities for the two classes given the observed feature vector, our decision would be easy. That is, if we knew $P(0|\overline{x})$ and $P(1|\overline{x})$, we would be in a situation similar to the featureless case. The best decision would simply be to choose the class with larger conditional probability, that is to say, decide 0 if $P(0|\overline{x}) > P(1|\overline{x})$ and decide 1 if $P(1|\overline{x}) > P(0|\overline{x})$. If $P(0|\overline{x}) = P(1|\overline{x})$, then they must both equal 1/2 and it does not matter what we decide.

The problem is to obtain $P(0|\overline{x})$ and $P(1|\overline{x})$. We have assumed that we know the prior probabilities $P(0)$ and $P(1)$ and we also know the conditional probabilities $P(\overline{x}|1)$ and $P(\overline{x}|0)$. Using these, we need to get the reverse conditional probabilities. A result from probability known as Bayes theorem is exactly what we need. We discuss this in Chapter 5.

4.5   CONTINUOUS FEATURES AND DENSITIES

As we mentioned previously, in many cases the features comprising the vector $\overline{x}$ are thought of as taking on a continuous range of values. Typically, each feature (each component of $\overline{x}$) is assumed to be a real-valued number. The feature vector is then a real-valued vector again taking on a continuum of values.

In such cases, instead of talking about the conditional probabilities $P(\overline{x}|0)$ and $P(\overline{x}|1)$, we need to allow conditional *densities* $p(\overline{x}|0)$ and $p(\overline{x}|1)$. These are just like the densities discussed in Chapter 3, but we have one for each class. Depending on whether the true object comes from class 0 or class 1, the feature vector $\overline{x}$ will be drawn according to the density $p(\overline{x}|0)$ or $p(\overline{x}|1)$, respectively. If the object belongs to class 0, then the probability that the feature vector $\overline{x}$ falls in some volume V is given by

$$P(\overline{x} \in V|0) = \int_V p(\overline{z}|0) \, d\overline{z}$$

as we would expect from our previous discussions of probability densities. The variable $\overline{z}$ is just a dummy variable of integration.

Although the feature vector takes on a continuum of values (necessitating the use of densities instead of probabilities), we still have only two classes 0 and 1. Thus, we still use prior probabilities $P(0)$ and $P(1)$ for the two classes. Furthermore, after observing the feature vector $\overline{x}$, we *still* have only two classes, so we can still talk about the conditional probabilities $P(0|\overline{x})$ and $P(1|\overline{x})$ as well. There is only a slight difference in how these conditional probabilities are computed, as we discuss in Chapter 5.

4.6 SUMMARY

In this chapter we discussed a probabilistic formulation for the pattern recognition problem. We assume that there are prior probabilities for the two classes and conditional probabilities for the feature vector, given each of the two classes. Under these distributions, we can compute the probability of error for any decision rule. We seek decision rules that minimize the probability of error. We suggested that the best decision rule is one that chooses the class with the larger posterior probability, given the observed feature vector. However, we did not discuss how to compute these posterior probabilities. This is done in the Chapter 5 along with a discussion of a useful result from probability called *Bayes theorem*. Finally, we ended the chapter with a discussion of the case in which the feature vector takes on a continuum of values (rather than a discrete set of values). In this case, the main difference is that conditional densities are used in the formulation instead of conditional probabilities.

4.7 APPENDIX: UNCOUNTABLY MANY

Natural numbers are potentially used to count things: $1, 2, 3, \ldots, n, \ldots$. There are infinitely many of these natural numbers, because for any finite natural number n, there is a larger number $n + 1$.

We can say that there are *countably many* members of a set F if each element of F can be assigned a unique natural number starting with 1 and continuing in order. Such an assignment provides an *enumeration* of F. If there is a greatest natural number n to which a member of F is assigned, then there are finitely many Fs and the number of Fs is n. If there is no such greatest number n, so that there are infinitely many Fs, but each element of F is assigned a unique natural number, then we say that F is *countably infinite* or *countable* for short.

Such a mapping of the elements of F to the natural numbers is equivalent to an ordering of all the members of the set, since the "order" will be given by the member that gets mapped to 1, then the member that gets mapped to 2, then to 3, etc. Clearly, by this definition, the natural numbers themselves are countable. So are the even numbers, because they can be put in the order 2, 4, 6, ... and then associated with the natural numbers in the order 1, 2, 3, ...

The set of all integers $\ldots, -3, -2, -1, 0, +1, +2, +3, \ldots$ is countable because they can be ordered as follows: $0, +1, -1, +2, -2, +3, -3, \cdots$ and then can be associated with the natural numbers in order, 0 with 1, $+1$ with 2, -1 with 3, etc.

Another example is the set of *rational* numbers. A rational number is a fraction with an integer numerator and a positive integer denominator: $\frac{m}{n}$. Examples include $\frac{1}{3}$, $\frac{-9}{7}$, and $\frac{2}{4}$. Any fraction is or has a unique reduction to a fraction in which the numerator and denominator have no common divisor greater than 1. For example, the fraction $\frac{2}{4}$ can be reduced to $\frac{1}{2}$.

Between any two different rational numbers there is always another rational number, so we cannot say that one rational number is right next to another. Nevertheless, the rational numbers between 0 and 1 are countable, because we can order them using their unique reductions in a way that allows them to be mapped in a one-to-one manner onto the natural numbers: $(\frac{1}{2}, \frac{1}{3}, \frac{2}{3}, \frac{1}{4}, \frac{3}{4}, \frac{1}{5}, \cdots)$.

A set is said to have uncountably many members if each element of F can be distinguished from every other and there are not countably many members of F. In this case, we say that the set F is *uncountably infinite*, or *uncountable* for short. Table 4.1.

Table 4.1   Diagonal Construction

String 1:	0	1	0	1	0	1	0	...
String 2:	1	1	0	1	1	0	1	...
String 3:	1	1	1	1	0	0	0	...
String 4:	0	0	1	0	0	0	0	...
String 5:	1	0	1	0	1	1	0	...
String 6:	1	1	1	1	1	0	1	...
String 7:	0	0	0	1	1	1	1	...
⋮	⋮	⋮	⋮	⋮	⋮	⋮	⋮	⋱
New string:	1	0	0	1	0	1	0	...

For example, there are uncountably many infinite strings of 0s and 1s. This can be shown through a *diagonal argument*. Given any enumeration of infinite strings of 0s and 1s, a new string can be found that is not included in the enumeration. This new string begins with 0 if the first string in the enumeration begins with 1 and begins with 1 if the first string in the enumeration begins with 0. The second item in the new string is 0 if the second item in the second string in the enumeration is 1 and vice versa. More generally, the nth item in the new string is 0 if the nth item in the nth string in the enumeration is 1, and otherwise the nth item in the new string is 1. For any n, the new string differs from the nth string at the nth place. So the new string cannot be included in the enumeration.

A similar diagonal argument can be used to show that the set of real numbers is uncountable, because any real number can be represented by an infinite decimal expansion.

4.8   QUESTIONS

1. What is a decision rule? If there are 10 possible feature vectors, how many possible decision rules are there?

2. What are the advantages of choosing/using more features and what are the disadvantages?

3. If the problem is to classify an email as spam or not spam, what features might you consider?

4. (a) As discussed in Chapter 1, a very common choice for object representation is a vector of real numbers. In some cases, the entries of the vector take on a discrete set (i.e., finite number) of values. Describe a situation in which having a discrete set of values is natural. (b) In implementing algorithms on a digital computer, inevitably we must eventually work only with discrete values. If so, why do we consider models in which the features are allowed to take on all real values?

5. Define what is meant by the probability of error of a decision rule. For any particular learning problem, what is the smallest probability of error of a decision rule? Is the best decision rule the one with the smallest probability of error?

6. If Ω_0 and Ω_1 are the subsets for which a decision rule c decides 0 and 1, respectively, write the expression for the average error rate of c?

7. In formulating the pattern recognition problem, we consider only the probability of error as our criterion for success. In some cases, this is not such an appropriate criterion since certain errors may be more costly than others. Describe one such situation, and suggest how one might modify the success criterion to address this issue.

8. Assuming that a decision rule for a pattern recognition problem is as follows: decide 1 (with cost \$10 for a wrong answer) when $0 \leq x \leq 1$, and decide 0 (with cost \$5 for a wrong answer) when $1 < x \leq 2$ and the feature space is [0,2], with no cost for a right answer. What is the expression (in terms of $P(0)$, $P(1)$, $p(x|0)$ and $p(x|1)$) for the average cost of that rule?

9. Are there uncountably many rational numbers?

4.9 REFERENCES

Much of the growth of statistical pattern recognition as a distinct discipline began in the 1960s. Some early survey articles and textbooks include Nilsson (1965); Nagy (1968); Ho and Agrawala (1968); Watanabe (1969), and Bongard (1970). There are many more recent books and articles on these subjects. See, for example, Devijver and Kittler (1982); Fukunaga (1990); and Schalkoff (1992). Duda *et al.* (2001) provide a readable treatment, which updates the now classic first edition of this book by Duda and Hart in (1973). Devroye *et al.* (1996) give an excellent and in-depth treatment. Kulkarni *et al.* (1998) provide a survey.

Bongard M. Pattern recognition. Washington (DC): Spartan Books; 1970.

Devijver PR, Kittler J. Pattern recognition: a statistical approach. Englewood Cliffs (NJ): Prentice-Hall; 1982.

Devroye L, Györfi L, Lugosi G. A probabilistic theory of pattern recognition. New York: Springer Verlag; 1996.

Duda RO, Hart PE, Stork DG. Pattern classification. 2nd ed. New York: Wiley; 2001.

Fukunaga K. Introduction to statistical pattern recognition. 2nd ed. San Diego (CA): Academic Press; 1990.

Ho YC, Agrawala A. On pattern classification algorithms: introduction and survey. Proc IEEE 1968;56:2101–2114.

Kulkarni SR, Lugosi G, Venkatesh S. Learning pattern classification—A survey. IEEE Trans Inf Theory 1998;44(6):2178–2206.

Nagy G. State of the art in pattern recognition. Proc IEEE 1968;56:836–862.

Nilsson NJ. Learning machines. New York: McGraw-Hill; 1965.

Schalkoff RJ. Pattern recognition: statistical, structural, and neural approaches. New York: Wiley; 1992.

Watanabe MS. Knowing and guessing. New York: Wiley; 1969.

CHAPTER 5

The Optimal Bayes Decision Rule

In the previous chapter, we formulated the following pattern recognition problem. There are two classes 0 and 1 with prior probabilities $P(0)$ and $P(1)$. We observe a vector of features $\overline{x}$ of the object that is related to the class of the object through conditional probabilities $P(\overline{x}|0)$ and $P(\overline{x}|1)$ (in the case of discrete-valued features) or conditional densities $p(\overline{x}|0)$ and $p(\overline{x}|1)$ (in case of continuous-valued features). We wish to decide whether the object belongs to class 0 or class 1, and we would like a decision rule that minimizes the probability of error.

With no observations, the obvious way to minimize the probability of error is to choose the class with larger prior probability. Now we would like to find out how to incorporate the fact that we get to observe the feature vector $\overline{x}$. A natural idea is to try to compute the conditional probabilities of the classes, given that we have observed the feature vector $\overline{x}$, namely $P(0|\overline{x})$ and $P(1|\overline{x})$. These give our "updated" assessment of the probability that the unknown object belongs to each of the two classes *after* observing the feature vector $\overline{x}$. If we had these probabilities, then the obvious way to try to minimize the probability of error would be to decide the class with larger conditional probability. But how can we compute the conditional probabilities that we need?

5.1 BAYES THEOREM

Recall the notion of the conditional probability of an event (Chapter 2). If A and B are two events with $P(A) > 0$, then the conditional probability of B given A, denoted by $P(B|A)$, is defined as

$$P(B|A) = \frac{P(B\&A)}{P(A)}. \qquad (5.1)$$

An Elementary Introduction to Statistical Learning Theory, First Edition.
Sanjeev Kulkarni and Gilbert Harman.
© 2011 John Wiley & Sons, Inc. Published 2011 by John Wiley & Sons, Inc.

$P(B|A)$ measures how often the event B is likely to occur when we restrict attention to those instances in which A occurs. This notion of conditional probability is precisely what we have been using in relating the feature to the class of the object. That is to say, $P(x|0)$ measures how often the observed feature is likely to take on the value x when the object belongs to class 0.

Bayes theorem is an elementary but powerful result that lets us switch the order of the events involved in the computation of conditional probability. That is, Bayes theorem lets us determine $P(A|B)$ in terms of $P(B|A)$ and other quantities.

The derivation of Bayes theorem is quite simple. From Equation (5.1) we multiply both sides by $P(A)$ to get

$$P(B\&A) = P(A)P(B|A). \tag{5.2}$$

Incidentally, recall that events A and B are independent if the occurrence of one does not affect the probability of the other—in other words, $P(B|A) = P(B)$. In this case, we get the familiar result that $P(B\&A) = P(B)P(A)$.

Now, assuming $P(B) > 0$, we can also use the definition of conditional probability to get an expression for $P(A|B)$ (just as we did for $P(B|A)$). We get

$$P(A|B) = \frac{P(B\&A)}{P(B)}, \tag{5.3}$$

where we have used the trivial fact that $P(A\&B) = P(B\&A)$. Multiplying both sides of Equation (5.3) by $P(B)$, we get

$$P(B\&A) = P(B)P(A|B). \tag{5.4}$$

In Equations (5.2) and (5.4), we now have two expressions for $P(B\&A)$, which of course must be equal. Therefore,

$$P(B)P(A|B) = P(A)P(B|A),$$

and dividing both sides by $P(B)$, we get Bayes theorem.

Bayes Theorem   If $P(A) > 0$ and $P(B) > 0$, then

$$P(A|B) = \frac{P(A)P(B|A)}{P(B)}.$$

5.2 BAYES DECISION RULE

The significance of Bayes theorem to the pattern recognition problem is that we can use the conditional probability of the evidence given the hypothesis to calculate the conditional probability of the hypothesis given the evidence, as long as we know the a priori probabilities of the hypothesis and the evidence.

We are assuming that the prior probabilities $P(0)$ and $P(1)$ and the conditional probabilities $P(\overline{x}|0)$ and $P(\overline{x}|1)$ are all known. If we could compute $P(\overline{x})$, then we could use Bayes theorem to obtain $P(0|\overline{x})$ and $P(1|\overline{x})$, which are the probabilities of the two classes, given our observation.

The unconditional probability $P(\overline{x})$ is given by

$$P(\overline{x}) = P(0)P(\overline{x}|0) + P(1)P(\overline{x}|1). \tag{5.5}$$

All the quantities on the right hand side are assumed to be known. The probabilities $P(0|\overline{x})$ and $P(1|\overline{x})$ are then given by

$$P(0|\overline{x}) = \frac{P(0)\,P(\overline{x}|0)}{P(\overline{x})} \tag{5.6}$$

and

$$P(1|\overline{x}) = \frac{P(1)\,P(\overline{x}|1)}{P(\overline{x})}. \tag{5.7}$$

In analogy with the featureless case, we expect that the best decision is simply to choose the class with larger conditional probability, that is, decide 0 if $P(0|\overline{x}) > P(1|\overline{x})$ and decide 1 if $P(1|\overline{x}) > P(0|\overline{x})$. If $P(0|\overline{x}) = P(1|\overline{x})$, then they must both equal 1/2 and it does not matter what we decide.

This is called *Bayes decision rule*, and it is in fact the optimal decision rule in the sense that no other decision rule has a smaller probability of error.

5.3 OPTIMALITY AND SOME COMMENTS

The probabilities $P(0|\overline{x})$ and $P(1|\overline{x})$ are often called *the posterior probabilities* of the two classes. The term "posterior" refers to the fact that these are probabilities obtained *after* the data have been taken into account, in contrast to the prior probabilities, which are obtained *before* any data have been taken into account. The posterior probabilities can be thought of as updating the prior probabilities, given the observed data.

To see that the Bayes decision rule is optimal, suppose that we observe the feature vector $\overline{x}$. A decision rule c must decide either 0 or 1, given the observation $\overline{x}$. If c decides 0 after observing $\overline{x}$, then it makes an error with probability $P(1|\overline{x})$, which is the probability that the object belongs to class 1 given the data, that is, the posterior probability of class 1. Likewise, if c decides 1, then it makes an error with probability $P(0|\overline{x})$. Once the value $\overline{x}$ is observed, the error can be minimized by deciding the class with the larger posterior probability.

The above argument holds for every possible value of the feature vector $\overline{x}$. That is, whatever feature vector is observed, the best decision (based on observing that particular feature value) is to decide that class with the larger posterior probability. But this is precisely Bayes decision rule!

Of course, if the two posterior probabilities are the same (and hence both equal to 1/2), then it does not matter which class we decide. The probability of error for either decision is 1/2, since after observing the data both classes are equally likely. Thus, there is not necessarily a *single* "best" rule—a number of rules may all have the same probability of error. However, no rule can perform strictly better than Bayes decision rule.

In Bayes decision rule, the denominators of the expressions for $P(0|\overline{x})$ and $P(1|\overline{x})$ are the same, namely $P(\overline{x})$. So, to decide which is larger, we do not actually need to compute $P(\overline{x})$. Instead, we can simply compare $P(0)P(\overline{x}|0)$ and $P(1)P(\overline{x}|1)$ and make a decision on the basis of which of these is larger.

However, to compute the error rate of Bayes rule, we typically do need $P(\overline{x})$. Bayes rule decides the class corresponding to the *larger* of $P(0|\overline{x})$ and $P(1|\overline{x})$. Therefore, the larger of these is exactly the probability that our decision is correct. This means that the probability of error, given we have observed $\overline{x}$, is the *smaller* of $P(0|\overline{x})$ and $P(1|\overline{x})$, or equivalently $\min\{P(0|\overline{x}), P(1|\overline{x})\}$. Thus, the overall (unconditional) probability of error of the Bayes decision rule can be expressed as

$$\text{Bayes error rate} = \sum_{\overline{x}} P(\overline{x}) \min\{P(0|\overline{x}), P(1|\overline{x})\}. \tag{5.8}$$

This is the best error rate that can be achieved. We denote this best possible error rate by R^*, so that

$$R^* = \text{Bayes error rate} = \sum_{\overline{x}} P(\overline{x}) \min\{P(0|\overline{x}), P(1|\overline{x})\}. \tag{5.9}$$

What can we say about R^*? Certainly, the error rate must satisfy $R^* \geq 0$, since the probability of error must be non-negative. Also, we could always ignore the feature vector and just randomly decide 0 or 1, each with a probability of 1/2. This simple rule has an error rate of 1/2, regardless of the underlying distributions. Hence, the Bayes error rate must satisfy the condition $R^* \leq 1/2$. We can also see this by noting that in Equation (5.8), either $P(0|\overline{x}) \leq 1/2$ or $P(1|\overline{x}) \leq 1/2$ since they must add up to 1. Therefore, $\min\{P(0|\overline{x}), P(1|\overline{x})\} \leq 1/2$, so that from Equation (5.8), we have

$$R^* = \sum_{\overline{x}} P(\overline{x}) \min\{P(0|\overline{x}), P(1|\overline{x})\}$$

$$\leq (1/2) \sum_{\overline{x}} P(\overline{x})$$

$$= 1/2.$$

Combining the lower and upper bounds, we have

$$0 \leq R^* \leq 1/2. \tag{5.10}$$

Without additional assumptions, we cannot say anything stronger, because there are cases that achieve the lower bound in Equation (5.10) and other cases that achieve the upper bound.

5.4 AN EXAMPLE

Suppose that we observationally determine that the unconditional probability of getting a 1 is 0.56 and the unconditional probability of getting a 0 is 0.44. We observe a single feature with six possible values, $v_1, \ldots, v_6$. For each possible value v_i, we determine the frequency with which items from class 0 have that value of the feature and the frequency with which items from class 1 have it. (Table 5.1)

We first compute the unconditional probabilities $P(v_i)$ for each feature i as

$$P(v_i) = P(1)P(v_i|1) + P(0)P(v_i|0).$$

For example,

$$P(v_1) = (0.56)(0.14) + (0.44)(0.05) = 0.0784 + 0.022 = 0.1004.$$

Now, suppose we observe v_1. Bayes theorem can be used to calculate the posterior probabilities $P(1|v_1)$ and $P(0|v_1)$.

$$P(1|v_1) = \frac{P(v_1|1)P(1)}{P(v_1)} = \frac{(0.14)(0.56)}{(0.1004)} \approx 0.78$$

and

$$P(0|v_1) = \frac{P(v_1|0)P(0)}{P(v_1)} = \frac{(0.05)(0.44)}{(0.1004)} \approx 0.22.$$

Alternatively, once we have $P(1|v_1)$, we can obtain $P(0|v_1)$ as

$$P(0|v_1) = 1 - P(1|v_1) \approx 1 - 0.78 = 0.22.$$

In this case, given that we observe v_1, the Bayes decision rule decides that the object belongs to class 1 (since $P(1|v_1) > P(0|v_1)$), and the conditional probability of error, given v_1, is $\min\{P(0|\overline{x}), P(1|\overline{x})\} \approx 0.22$.

Table 5.1 Observed Feature Frequencies

	v_1	v_2	v_3	v_4	v_5	v_6
$P(v_i|1)$	0.14	0.29	0.36	0.14	0.07	0.00
$P(v_i|0)$	0.05	0.09	0.23	0.27	0.25	0.11

Table 5.2   Conditional Probabilities of 1 and 0, given V_i

	v_1	v_2	v_3	v_4	v_5	v_6
$P(1|v_i)$	0.78*	0.8*	0.67*	0.4	0.27	0.00
$P(0|v_i)$	0.22	0.2	0.33	0.6*	0.73*	1.0*

Table 5.3   Using Different $P(1)$ and $P(0)$

	v_1	v_2	v_3	v_4	v_5	v_6
$P(1|v_i)$	0.96*	0.97*	0.93*	0.82*	0.72*	0.00
$P(0|v_i)$	0.04	0.03	0.07	0.18	0.28	1.0*

In a similar manner, we can find $P(0|v_i)$ and $P(1|v_i)$ for $i = 2, 3, 4, 5, 6$ as shown in Table 5.2. For each i, the Bayes decision rule is indicated by the asterisks. Thus, the optimal decision rule is to decide class 1 if we observe v_1, v_2, or v_3, and to decide class 0 if we observe v_4, v_5, or v_6.

The Bayes error rate can be computed using Equation 5.8, and is found to be 0.28.

Now suppose we have the same conditional probabilities as in the table above, but the prior probabilities of the two classes are $P(0) = 0.1$ and $P(1) = 0.9$. Then the posterior probabilities $P(0|v_i)$ and $P(1|v_i)$ for $i = 1, 2, 3, 4, 5, 6$ are as shown in Table 5.3, again with Bayes decision rule indicated by asterisks. In this case, we see that the optimal rule is to decide class 1 except when we observe the feature value v_6, in which case we decide class 0.

We see that in Bayes decision rule both the prior probability of a class as well as the conditional probability of seeing a feature, given the class, are important in determining the optimal decision. In the first example, the prior probabilities are about the same, so that the decision is largely governed by the probabilities of seeing the various feature values for each class. In the second example, the probability of seeing an object from class 0 is very low, so we tend to favor deciding class 1 as we would expect. However, if we see the feature value v_6, then it is so unlikely to have come from class 1 (in fact, probability 0 in this case) that this overcomes our prior bias and the best decision is class 0. Bayes decision rule tells us precisely how to balance our prior knowledge of the class probabilities with the new information provided by observing the feature.

5.5   BAYES THEOREM AND DECISION RULE WITH DENSITIES

It turns out that Bayes theorem and the resulting Bayes decision rule also work with densities in a completely analogous way. Suppose the features in the feature vector $\overline{x}$ are continuous valued, so that we have conditional densities $p(\overline{x}|0)$ and $p(\overline{x}|1)$ for the distributions of the feature vector for each class. The unconditional density of the feature vector is given by

$$p(\overline{x}) = P(0)p(\overline{x}|0) + P(1)p(\overline{x}|1).$$

If the density $p(\overline{x}) > 0$ and $P(0)$ and $P(1) > 0$, then Bayes theorem for densities tells us that

$$P(0|\overline{x}) = \frac{P(0)\ p(\overline{x}|0)}{p(\overline{x})}$$

and

$$P(1|\overline{x}) = \frac{P(1)\ p(\overline{x}|1)}{p(\overline{x})}.$$

This looks just like the expressions we had before in Equation 5.6, except that the capital "P" denoting probabilities for $\overline{x}$ have now been replaced by lower case "p" denoting the density for $\overline{x}$. But since the class of the object is still discrete (namely, just two classes—either 0 or 1), we still have probabilities for the class of the object.

Bayes decision rule is exactly the same as in the purely discrete case. That is to say, we decide class 0 if $P(0|\overline{x}) > P(1|\overline{x})$ and we decide class 1 if $P(1|\overline{x}) > P(0|\overline{x})$. If $P(0|\overline{x}) = P(1|\overline{x})$ then both are equal to 1/2 and, as before, our decision will not affect the probability of error.

Given that we see a particular feature vector $\overline{x}$, our conditional probability of error is again $\min\{P(0|\overline{x}), P(1|\overline{x})\}$. The overall (unconditional) probability of error is therefore

$$R^* = \text{Bayes error rate} = \int \min\{P(0|\overline{x}), P(1|\overline{x})\}\, p(\overline{x})\, \mathrm{d}\overline{x}. \qquad (5.11)$$

This is very similar to the expression in Equation 5.8 except that now the probability $P(\overline{x})$ is replaced by the density $p(\overline{x})$, and therefore, the summation over discrete feature vectors is replaced by an integral. Again, this is the best error rate that can be achieved over all possible decision rules.

5.6 SUMMARY

In Chapter 4, we noted that to minimize the probability of error, we need to decide the class that has the larger posterior probability, given the observed feature vector. However, in that chapter we did not show how to compute the needed posterior probabilities. In this chapter we began with a result from probability known as *Bayes Theorem*. This result shows how to switch the order of events in a conditional probability, and this is exactly the tool that allows us to compute the needed posterior probabilities. The resulting decision rule is called *Bayes decision rule*. We argued that this is the optimal decision rule in the sense that no other rule can have a smaller probability of error. We discussed a simple example in detail. We ended with a discussion of Bayes Theorem and Bayes decision rule in the case of densities.

5.7 APPENDIX: DEFINING CONDITIONAL PROBABILITY

It is relatively standard practice to define conditional probability in terms of (unconditional) probability. This is actually controversial and some theorists have argued that the reverse is true; unconditional probability should be defined in terms of conditional probability (Hájek, 2003).

Standard accounts follow Kolmogorov (1933) in defining the conditional probability of B given A, $P(B|A)$, as follows:

$$P(B|A) = \frac{P(B\&A)}{P(A)}. \tag{5.12}$$

Since division by zero is not defined, standard accounts take conditional probability to be similarly undefined when $P(A) = 0$.

On the other hand, we have also suggested that $P(B|A)$ measures how often the event B is likely to occur when we restrict attention to those instances in which A occurs. Of course, if an event whose probability is 0 could not occur, this would not matter.

However, there are cases in which $P(A) = 0$ but A can still occur. Situations in which it is appropriate to appeal to *probability density* are almost always examples of this sort.

Suppose that Alice throws a dart at a circular dart board that is exactly 1 ft in diameter and suppose also that, if she hits the board, there is a particular point on the board that she hits. (We are concerned with the first two times in which she actually hits points on the board.) Suppose that she is equally likely to hit any of the points on the board and, in particular, that the probability density of her hitting a particular point on the board is constant over the whole dart board. Let A be that Alice's first dart hits a point whose distance from the center is exactly 3 inches. Let B be that Alice's second dart hits a point that is nearer to the center than her first dart.

Even though $P(A) = 0$, it is natural to think that $P(B|A)$ is defined in this case and (because of the uniform probability density) it is equal to the fraction of the area of the dart board that is within 3 inches of the center divided by the total area of the dart board: $P(B|A) = 1/4$.

So it seems that the conditional probability of B, given A, $P(B|A)$, can sometimes have a definite value even though $P(A)$ is 0. If so, we cannot use the standard definition of $P(B|A)$. And perhaps there is no way to define conditional probability in terms of unconditional probabilities.

On the other hand, it may be possible to define the unconditional probability of A, $P(A)$, in terms of the conditional probability of A, given something T that must always be certainly true. $P(A) = P(A|T)$. What would be a good choice for such a T?

5.8 QUESTIONS

1. What is Bayes *theorem*? Why is it true? How might Bayes theorem be useful in determining the probability of a hypothesis, given some evidence?

2. What is a Bayes rule?

3. True or false? If the features, classifications, and probabilities are fixed, there can be more than one Bayes rule.

4. Consider two classes 0 and 1 with $P(0) = 0.6$. Suppose that the conditional probability density functions of the feature x are $p(x|0) = 1$ for $0 \leq x \leq 1$ and $p(x|1) = 2x$ for $0 \leq x \leq 1$.

   (a) What is the feature space?
   (b) What are the possible decision rules?
   (c) What is the optimal (Bayes) decision rule? Draw a picture indicating the intervals where the decision is one and where the decision is zero.
   (d) What is the minimum (Bayes) error rate?

5. For the situation in question 4, consider two decision makers, Ivan and Derek, who make a decision solely on the basis of prior probabilities (i.e., without observing any features). Ivan thinks that the decision should be randomized according to the prior probability distribution. So, he decides "0" 60% of the times and "1" 40% of the times, on the basis of the outcome of a biased coin flip. Derek, on the other hand, thinks that he should always decide the greater of the two probabilities, that is, "0" all the times. Who do you think will perform better? What are their error rates?

6. Consider the situation in question 4 and assume now that it costs $12 if you decide 0 when it really was a 1 and $10 if you decide 1 when it really was a 0.

   (a) If you use the Bayes decision rule that you derived for question 4, how much does it cost you to make a decision on average?
   (b) Imagine that you observe the feature $x = 0.7$. If you decide that the class was 0, what is your average cost? If you decide that the class is 1, what is your average cost? Which should you decide to minimize costs?

7. Continuing with the cost assumptions of the previous problem.

   (a) In general, when you observe a feature x, what is the average cost if you decide 0? What is it if you decide 1? (These should be dependent on x.)
   (b) On the basis of these average costs, to minimize the average cost, when should you decide 0? When should you decide 1?
   (c) If you use this decision rule, what is your average cost of making a decision?

8. There are two containers of black and white balls. In container 1, 1/3 of the balls are black and 2/3 are white. In container 2, 2/3 of the balls are black and 1/3 are white. The container before you is equally likely to be container 1 or container 2. You experiment by withdrawing a ball, checking its color, and returning it. Given that the ball is white, what is the probability that the container before you is container 1? If you repeat the experiment three times, What is the probability of getting the result white-black-white?

9. Suppose there are two curtains and behind each curtain is either gold or a goat. Suppose the prior probability that there is gold behind a given curtain is 1/2, is the same for each of the curtains, and is independent of what is behind the other curtain.

    (a) If you are told that there is gold behind Curtain 1, what is the posterior probability that there is gold behind both curtains, given that information?

    (b) If you are told that there is gold behind at least one of the curtains, what is the posterior probability that there is gold behind both curtains, given that information?

10. Suppose we know that the probability that it will rain on any given day is p and is independent of the weather on all other days. As a meteorologist, what should your pattern of weather predictions be from day to day to minimize your probability of error, and what is the resulting error rate? Suppose as your prediction for tomorrow's weather you use the weather you observe today. What is the error rate for this prediction rule?

11. A die is weighted so that when it is tossed, one particular number will come up with a probability of 1/2 and each of the five other numbers will come up with a probability of 1/10. Suppose that each of the six sides has had an equal chance of being the side that is favored. We are interested in deciding which is the favored number.

    (a) With no observations, what is an optimal decision rule and what is the probability of error?

    (b) Suppose we observe a 3 and a 4 on two independent rolls of the die. Given this data, what is an optimal decision rule and what is the probability of error?

12. Suppose that 5% of women and 0.25% of men in a given population are colorblind.

    (a) Write Bayes Theorem.

    (b) If a colorblind person is chosen at random from a population containing an equal number of males and females, what is the probability that the chosen colorblind person is male?

Let c denote the event that the selected person is colorblind, m denote that the selected person is male, and f denote the selected person is female.

The probability that a randomly selected person is colorblind is then $P(c) = P(c|m)P(m) + P(c|f)P(f) = (1/20)*(1/2) + (1/400)*(1/2) = 21/800$.

So $P(m|c) = P(c|m)P(m)/P(c) = (1/400)*(1/2)/(21/800) = 1/21$.

(c) If the population has twice as many males as females, what is the probability that a colorblind person chosen at random is male?

$P(m) = \frac{2}{3}$, $P(f) = \frac{1}{3}$. Then $P(c) = P(c|m)P(m) + P(c|f)P(f) = (1/400)(2/3) + (1/20)(1/3) = 11/600$.

So $P(m|c) = P(c|m)P(m)/P(c) = (1/400)(2/3)/(11/600) = 1/11$.

(d) If the population has an equal number of males and females, what is the probability that a person chosen at random is both female and colorblind?
$P(c \text{ and } f) = P(c|f)P(f) = (1/20)*(1/2) = 1/40$.

(e) If the population has twice as many males as females, what is the probability that a person chosen at random is both male and not colorblind?
$P(\text{not } c \text{ and } m) = (1 - P(c|m))P(m) = (1 - 1/400)*(2/3) = 133/200$.

13. Suppose the feature vector x can only take values -1 or $+1$, and the Bayes rule is as follows: decide 0 when $x = -1$ and decide 1 when $x = +1$. Suppose the cost of a correct decision is 0, but the cost of an incorrect decision is as follows. The cost of deciding 0 when actual class is 1 is \$5 and the cost of deciding 1 when the class is 0 is \$10. What is the expression for the average cost of the decision rule (in terms of $P(x|0)$, $P(x|1)$, $P(0|x)$, $P(1|x)$, $P(0)$, and $P(1)$)?

14. True or false? $P(A|B)$ might have a definite value even if $P(B) = 0$.

5.9 REFERENCES

Bayes theorem (sometimes called *Bayes formula*) is an old result obtained by an eighteenth century mathematician Rev. Thomas Bayes (1763) and later extended by Laplace (1814). This theorem is covered in most introductory books on probability (see, e.g., any of the references on probability from Chapter 2). In particular, Everitt (1999) provides a readable treatment.

The use of Bayes theorem in statistics is widespread and has a long history. Its importance in statistical pattern recognition was recognized early on. When all relevant quantities are known, the resulting Bayes decision rule provides the statistically optimal decision (a benchmark) against which other methods (like those that we will discuss later) can be compared. Almost all books or surveys on statistical pattern recognition have an early chapter or section on Bayes decision rule. See, for example, Chapter 2 of Duda *et al*. (2001), Chapter 2 of Devroye *et al*. (1996), the introduction of Kulkarni *et al*. (1998), as well as the other pattern recognition references from Chapter 4.

Devroye L, Györfi L, Lugosi G. A probabilistic theory of pattern recognition. New York: Springer Verlag; 1996.

Duda RO, Hart PE, Stork DG. Pattern classification. 2nd ed. New York: Wiley; 2001.

Everitt BS. Chance rules: an informal guide to probability, risk, and statistics. New York: Copernicus, Springer-Verlag; 1999, Chapter 7.

Hájek A. What conditional probability could not be. Synthese 2003;137:273–323.

Kolmogorov AN. Grundbegriffe der wahrscheinlichkeitrechnung, ergebnisse der mathematik; translated as foundations of probability. New York: Chelsea Publishing Company; 1933.

Kulkarni SR, Lugosi G, Venkatesh S. Learning pattern classification—A survey. IEEE Trans Inf Theory 1998;44(6):2178–2206.

CHAPTER 6

Learning from Examples

In Chapters 4 and 5, we described the pattern classification problem and discussed the best possible decision rules, the Bayes decision rules. Although the Bayes decision rules will generally have a nonzero error rate (because of inherent randomness), they will do better than any other rules. If we can find a Bayes decision rule, we are guaranteed to have as small an error rate as possible.

However, in order to find such a decision rule, we need to know the underlying distributions $P(0)$, $P(1)$, $P(\overline{x}|0)$, and $P(\overline{x}|1)$, and we are rarely so lucky to have this knowledge. This is where learning from data comes in. Instead of assuming that we know the distributions governing the problem, let us assume that we have access to some prior examples or data. Let us try to use the data to learn to make good decisions by coming up with a relatively good decision rule.

This is a powerful idea, since in many applications we may know very little about the distributions, and yet we may be able to observe data that allow us to come up with good decision rules. Some immediate questions that arise are "How can we find good classification rules?," "What is the best performance we can expect?," and "Can we find rules that achieve this performance?" In this chapter we describe, in general terms, how to formulate this learning problem. In the rest of the book we address the questions above and discuss some of the most well-known and successful learning methods.

6.1 LACK OF KNOWLEDGE OF DISTRIBUTIONS

Recall the pattern recognition problem as we have discussed it so far. An object belongs to one of two classes, 0 and 1. We observe features of the object represented by a feature vector $\overline{x}$ and wish to decide whether the object belongs to class 0 or class 1. We would like to find a decision rule that minimizes the probability of error.

An Elementary Introduction to Statistical Learning Theory, First Edition.
Sanjeev Kulkarni and Gilbert Harman.
© 2011 John Wiley & Sons, Inc. Published 2011 by John Wiley & Sons, Inc.

In discussing the pattern recognition problem, we assumed a knowledge of the prior probabilities $P(0)$ and $P(1)$ that the object belongs to class 0 and 1 respectively, and the conditional probabilities $P(\overline{x}|0)$ and $P(\overline{x}|1)$ (or conditional densities $p(\overline{x}|0)$ and $p(\overline{x}|1)$ relating the observed feature vector to the class of the object. With this knowledge, given the feature $\overline{x}$, we can, in principle, implement a Bayes decision rule. That is, we can compute the posterior probabilities $P(0|\overline{x})$ and $P(1|\overline{x})$ and then decide that the object belongs to the class with the larger posterior probability. (If these posterior probabilities are the same, we can choose either class.)

A Bayes rule is a best decision rule in terms of minimizing the probability of error. Because of the randomness inherent in both the class to which the object belongs and the relationship between the observed features and the object's class, we should not expect any decision rule to work flawlessly. Even an optimal Bayes rule may make mistakes, that is, even an optimal rule may have some nonzero error rate. However, no other rule can do better.

Unfortunately, in many situations, we have no way actually to implement a Bayes rule. The problem is that we need to know $P(0)$, $P(1)$, $P(\overline{x}|0)$, and $P(\overline{x}|1)$, and we are rarely fortunate enough to have this knowledge. Of these quantities, $P(0)$ and $P(1)$ are easier to determine than the conditional distributions. The reason is that $P(0)$ is just a single number, namely the prior probability that an object comes from class 0, and $P(1)$ is just $1 - P(0)$. Even if we do not know these numbers, we might be able to come up with reasonable estimates fairly easily. However, the conditional distributions are functions of $\overline{x}$, which makes them much more difficult to estimate.

For example, consider the problem of handwritten character recognition. Suppose a machine can scan individual handwritten letters, resulting in a digital image of the letters, and we wish to recognize which letter of the alphabet is written. In this application, it is one thing to know (or assume that we know) the prior probabilities of each letter, but an altogether different thing to assume that we know the conditional probabilities of the feature vector, given the letter.

Indeed, the prior probability of each letter can be identified with the long-run frequency of occurrence of that letter, for which we may have a good estimate. On the other hand, consider the conditional distributions of the feature vector, given the letter, namely $P(\overline{x}|a)$, $P(\overline{x}|b), \ldots, P(\overline{x}|z)$. There are 26 of these (even ignoring lower case versus upper case letters, punctuation marks, etc.). The feature vector $\overline{x}$ in this case represents the digital image, and $P(\overline{x}|a)$ denotes the probability of seeing the particular image $\overline{x}$, given that the letter written is an "a." There are an enormous number of probabilities (one for each image and letter). We generally have very little knowledge of these and even trying to come up with good models or estimates is impractical.

6.2 TRAINING DATA

We now assume that, although nature is governed by some prior probabilities and conditional distributions, these are unknown to us. What can we do in this case to come up with good decision rules? It seems clear that, unless we have some additional information upon which to base a decision, there is not much that we can do besides randomly deciding on one or another of the classes.

A powerful approach—that is the subject of the rest of this book—is to assume that we have access to *training data* consisting of a set of labeled examples, $(\overline{x}_1, y_1), (\overline{x}_2, y_2), \ldots, (\overline{x}_n, y_n)$. The $\overline{x}_i$ are observed feature vectors that characterize certain objects and each y_i indicates the class to which object i belongs (either 0 or 1).

Given the training data, we then observe a new object characterized by the feature vector $\overline{x}$ and wish to decide whether the new object belongs to class 0 or class 1. This is the basic problem of *learning from examples*. It is commonly referred to as "supervised learning," since we have access to a set of examples that are assumed to be properly labeled by a "supervisor."

The information available to classify an object is given by the training data $(\overline{x}_1, y_1), (\overline{x}_2, y_2), \ldots, (\overline{x}_n, y_n)$ as well as the feature vector $\overline{x}$ of the new object to be classified. Given all of this information, our decision is either 0 or 1. Thus, a decision rule $c(\cdot)$ that we come up with will generally depend on all of this information so that $c(\overline{x}; (\overline{x}_1, y_1), (\overline{x}_2, y_2), \ldots, (\overline{x}_n, y_n))$ is either 0 or 1.

Often we suppress the explicit dependence of the decision rule on the training data. That is, given the training data $(\overline{x}_1, y_1), (\overline{x}_2, y_2), \ldots, (\overline{x}_n, y_n)$, our aim is to come up with a decision rule $c(\cdot)$ that decides 0 or 1, given a feature vector $\overline{x}$. This is exactly the kind of decision rule we discussed in Section 4.2. That is to say, the decision rule $c(\overline{x})$ can be thought of as a function $c : \mathbf{R}^d \rightarrow \{0, 1\}$ that takes as input the feature vector $\overline{x}$ of d real numbers, representing the values of the d features, and produces a decision $c(\overline{x})$ that is either 0 or 1.

Of course, the decision rule $c(\overline{x})$ is obtained using some learning algorithm on the training data. Thus, we can think of a learning algorithm as taking training data $(\overline{x}_1, y_1), (\overline{x}_2, y_2), \ldots, (\overline{x}_n, y_n)$ as input and producing a decision rule $c(\overline{x})$ as output. We would like to find good learning algorithms. That is, we would like to find methods that use the training data to come up with (or "learn") good decision rules.

The idea of incorporating learning into the pattern recognition problem is extremely powerful. If we are able to come up with good classification rules, we will have achieved a great deal since we will have replaced the requirement that the distributions be known with the requirement that we be given a set of labeled examples. In many applications, knowledge of the distributions is difficult to obtain but labeled examples are relatively easy to come by. Consider again the problem of handwritten character recognition. We have mentioned how hopeless it may be to try to obtain the conditional distributions for this problem. On the other hand, to obtain labeled examples, we merely have to select a set of handwritten characters and have a human supervisor classify these examples.

6.3 ASSUMPTIONS ON THE TRAINING DATA

We would like whatever decision rule we come up with to perform well. But what do we mean by this? We can use the same criterion as before—namely, the probability of error. That is, we will assume that there are prior probabilities and conditional probability distributions governing the problem, even though we do not know what they are. We would like to have a small probability of error according to these unknown distributions.

As a benchmark, we would like to compare the performance of whatever rule we arrive at to the Bayes error rate R^*. But, how can we compare the performance of our rule with R^* when we cannot even compute it since we do not know the underlying distributions? It turns out that there are results about how close we can get to the Bayes error rate R^*, whatever that error rate may be, even though it is unknown.

Of course, if we hope to come up with good decision rules on the basis of data, we need useful data. We need to assume that the data are *representative* of the new objects we will be asked to classify. Since we are measuring performance by a probability of error, given the underlying distributions, the labeled examples we observe should somehow reflect those distributions.

We will assume that the labeled examples $(\overline{x}_1, y_1), (\overline{x}_2, y_2), \ldots, (\overline{x}_n, y_n)$ arise from these same distributions. Furthermore, we will assume that results of these distributions are "i.i.d."—independent and identically distributed. This means that each example, including each labeled example used as data and any later example to be classified, arises from the same underlying probability distributions $P(0)$, $P(1)$, $P(\overline{x}|0)$, and $P(\overline{x}|1)$ (so the examples are identically distributed) and also that the feature vector and label of any example does not depend probabilistically on the feature vector and label of any other example (so the examples are probabilistically independent).

The reason for the "identically distributed" condition is straightforward. Consider again the character recognition problem. Suppose the training data contained examples of only the letters "A" and "B" and did not contain examples of the other letters, even though these other letters do show up in the samples we wish to learn to classify. How could we possibly expect to come up with a rule that classifies these other letters correctly? Or, to take a less extreme example, imagine that the training data contained examples of only neatly written letters, or letters in just one type of handwriting, but the samples we wish to classify do not contain only neatly written letters in that style. Then how could we expect our system to perform well on messily written characters, or in a different style of handwriting? To come up with a decision rule that works well according to the distributions $P(0)$, $P(1)$, $P(\overline{x}|0)$, and $P(\overline{x}|1)$, it makes sense that the training data should arise from these same distributions. If the problem is to find a rule to classify actual written characters on envelopes, for example, the training data should be taken from a random sample of actual writing on envelopes.

The "independence" assumption is somewhat more difficult to explain. Consider the following extreme case. Suppose our sampling procedure is to draw the first training example $(\overline{x}_1, y_1)$ randomly according to the underlying distributions $P(0)$, $P(1)$, $P(\overline{x}|0)$, and $P(\overline{x}|1)$, and the rest of the training examples are taken to be the same as $(\overline{x}_1, y_1)$. Rather than drawing the new training examples randomly, we just take $(\overline{x}_2, y_2) = (\overline{x}_1, y_1)$, $(\overline{x}_3, y_3) = (\overline{x}_1, y_1)$, and so on up to $(\overline{x}_n, y_n) = (\overline{x}_1, y_1)$.

Then we could argue that each example is drawn according to the distributions $P(0)$, $P(1)$, $P(\overline{x}|0)$, and $P(\overline{x}|1)$. So the training examples are identically distributed (in fact, they are identical!). However, there is clearly a problem here. We have no hope of coming up with a good decision rule based on just seeing $(\overline{x}_1, y_1)$ repeated many times. The problem is that the training examples are not independent—rather they are completely dependent, in effect giving us just one example.

One might guess that requiring the data to be strictly independent and identically distributed is not absolutely necessary for learning. The main requirement is that the data are rich and representative enough to give a good sense of the posterior distributions $P(\overline{x}|0)$ and $P(\overline{x}|1)$ for all $\overline{x}$, since this is what Bayes decision rule uses to classify $\overline{x}$. Indeed, the assumption that the training data are independent and identically distributed can be relaxed, and a good deal of work has been done on understanding more general conditions under which successful learning can take place. In particular, much work has been done on allowing some dependence in the training examples (e.g., see, Györfi *et al.* (1989) and references therein for the estimation problem). However, the i.i.d. assumption is the simplest to work with and yet it captures the richness needed to discuss learning methods useful for most applications.

6.4 A BRUTE FORCE APPROACH TO LEARNING

Faced with the learning problem and armed with the tools from Chapters 4 and 5, a natural approach suggests itself: we could try to use the data $(\overline{x}_1, y_1)$, $(\overline{x}_2, y_2), \ldots, (\overline{x}_n, y_n)$ to estimate the unknown distributions $P(0)$, $P(1)$, $P(\overline{x}|0)$, and $P(\overline{x}|1)$. Then we could use these estimates to construct a Bayes decision rule corresponding to the estimated distributions. The hope is that if the estimates are close to the underlying true distributions, then the resulting decision rule will be close to the true Bayes decision rule (that corresponds to the underlying distributions).

As we mentioned before, $P(0)$ and $P(1)$ are not hard to estimate. Since $P(0)$ is the prior probability that an object comes from class 0, a natural estimate, $\hat{P}(0)$, is to simply count the number of training examples that are labeled 0 and to divide this by n. Likewise, a natural estimate $\hat{P}(1)$ for $P(1)$ is the number of training examples labeled 1 divided by n. Of course, $\hat{P}(1) = 1 - \hat{P}(0)$, as expected.

However, the conditional distributions are functions of $\overline{x}$, and hence these are much more difficult to estimate. If the feature vector takes on only a finite set of values, say $\overline{x} \in \{a_1, a_2, a_3, a_4, a_5\}$, then the conditional distribution $P(\overline{x}|0)$ is represented by five numbers, namely $P(a_1|0), \ldots, P(a_5|0)$. $P(a_1|0)$ can be estimated by counting the number of training examples for which $y_i = 0$ and $\overline{x}_i = a_1$ and dividing by the number of training examples for which $y_i = 0$. $P(a_2|0), \ldots, P(a_5|0)$ can be estimated similarly. Likewise, $P(\overline{x}|1)$ can be estimated in an analogous way.

In the case that the feature vector $\overline{x}$ takes on a continuum of values, the conditional distributions are typically densities $p(\overline{x}|0)$ and $p(\overline{x}|1)$. There are methods for density estimation that in principle could be used to provide estimates $\hat{p}(\overline{x}|0)$ and $\hat{p}(\overline{x}|1)$ for the densities using the training data.

Once we have estimates $\hat{P}(0)$, $\hat{P}(1)$, $\hat{P}(\overline{x}|0)$, and $\hat{P}(\overline{x}|1)$, we can substitute these into Equations (5.5) and (5.6) to form an estimate of Bayes decision rule. The problem with this approach is that in most practical applications, the conditional distributions are extremely difficult to estimate. Typically, the amount of available training data is much too small to give reasonable estimates. And, trying to come up with good models for the conditional distributions on the basis of prior information about the problem is also very difficult.

Thus, the brute force approach is generally impractical as discussed further in the following section. Fortunately, to come up with decision rules, we do not need to estimate the underlying distributions, and in fact trying to do so is overkill. Recall that Bayes decision rule is based on the posterior probabilities $P(0|\overline{x})$ and $P(1|\overline{x})$. Moreover, Bayes rule is simply based on which of the posterior probabilities is larger, so we do not even need to know them exactly. Given the training data $(\overline{x}_1, y_1), (\overline{x}_2, y_2), \ldots, (\overline{x}_n, y_n)$ and a new feature vector $\overline{x}$, we need to only decide which of $P(0|\overline{x})$ and $P(1|\overline{x})$ is larger.

Moreover, if $P(0|\overline{x})$ and $P(1|\overline{x})$ are very different from each other (that is, one is close to 0 and the other is close to 1), then it might be easy to decide which is larger. If they are very close to each other (that is, both are close to 1/2), then a mistake in deciding which is larger does not contribute much to the error rate. So, there is hope even though the brute force approach is unlikely to work.

6.5   CURSE OF DIMENSIONALITY, INDUCTIVE BIAS, AND NO FREE LUNCH

A brute force approach works well if the number of possible values for $\overline{x}$ is small compared with the number of training examples, or, in the case of a continuum of values, if the number of dimensions is small. However, many practical applications do not satisfy these conditions. Even under modest assumptions, the number of dimensions can be large and the number of possible feature vectors, enormous.

Consider, first, a finite case but one for which the number of possible values for $\overline{x}$ is very large compared with the number of training examples. For example, suppose the feature vector is a rather small binary image, say just 10×10. The number of possible feature vectors is the number of 10×10 binary images. There are only 100 pixels and each pixel can take on only 2 possible values, but there are still 2^{100} possible feature vectors. With any reasonable amount of training data, it would be very difficult to use the discrete approach for estimating the conditional probabilities. If we considered grayscale images with 256 gray levels, then there would be 256^{100} possible feature vectors (images).

One might try to model the gray levels as taking on a continuum of values and use methods for density estimation. In this case, the number of dimensions is 100, which is very large as far as trying to get good density estimates is concerned. And in many practical applications, the number of dimensions can be much larger than 100. For example, with a more reasonable image size of say 100×100, the number of dimensions is 10,000, and even this is not a very high-resolution image.

In these and many other very practical cases, the brute force approach performs very poorly and becomes useless rapidly even for a modest number of dimensions. The rapid increase in difficulty of the learning problem as the number of dimensions increases is often referred to as the "curse of dimensionality."

The curse of dimensionality is very real, both in theory and practice. There are theorems that quantify various ways in which the difficulty of learning increases exponentially with the number of dimensions. (Notice, for example, that the volume of a d-dimensional hypercube with a side of length L is L^d.) Such results might seem to suggest that the learning problem is hopeless for many applications that have high-dimensional feature vectors with modest amounts of training data. Such problems are difficult, yet as we will see a number of practical learning algorithms provide results better than might be expected. How is this possible?

It turns out that in many applications, the underlying distributions have some structure so that the "effective" number of dimensions can be much smaller than the number of actual dimensions of the feature vector. If the particular learning method used somehow captures this structure, then learning with a reasonable number of training examples may be possible. On the other hand, if the structure favored by the learning method is ill-suited to the application at hand, then far more examples are needed for comparable performance.

The fact that a learning method favors some decision rules over others is called *inductive bias*, which is necessary for inductive learning. A completely unbiased system cannot learn from experience.

It would be nice if there were one learning method that outperformed all others regardless of the underlying distributions. Unfortunately, this is not the case. There are results quantifying this that are sometimes referred to as "no free lunch theorems." For this reason, a variety of techniques have emerged that have varying degrees of success, depending on the underlying problem. Of course, some methods have been shown to outperform others in many practical problems of interest. However, the no free lunch theorems guarantee that no single learning method can outperform all others on all problems, making the design of practical learning algorithms a mix of an art and science.

6.6 SUMMARY

In this chapter we discussed learning from examples. We assumed that there are prior probabilities and conditional probabilities as before, but we assumed that these are unknown. Knowledge of the distributions is replaced by a set of examples called training data. The problem of learning from examples is to use the training data to come up with a decision rule for classifying a new feature vector. Although the distributions are unknown, the performance of the decision rule learned is evaluated using the (unknown) distributions. In order to come up with a good rule, we need the training examples to be representative of the underlying distributions. The standard assumptions are that the training data are independent and identically distributed (i.i.d.) according to the underlying distributions. A brute force approach to learning from examples is to use the training data to estimate the unknown distributions and then construct a Bayes decision rule using our estimates. This is impractical in most applications because of the difficulty of estimating densities in high dimensions with minimal data. Though the methods we will be discussing avoid this problem to some extent (by directly finding classification rules rather than first estimating the distributions), learning from examples is inherently a difficult problem, especially in high dimensions. This is often called the curse of dimensionality. While it would be nice to have one learning method that beats all others, this is not possible.

6.7   APPENDIX: WHAT SORT OF LEARNING?

There are different kinds of learning. There is learning *that* something is the case, as when Jack learns that Betty is dating Doug. There is learning *how* to do something, as when Ken learns how to ride a bike. There is learning physics, learning Italian, learning tennis, and learning a poem.

Sometimes learning is an all or nothing matter, as in learning that Betty is dating Doug or learning a particular poem. Sometimes learning is a matter of degree: one gets better at physics or tennis, one learns more physics, one gets better at solving physics problems.

It is a debated question in philosophy and linguistics whether *knowing how* to do something is always a case of *knowing that* something is so (Stanley and Williamson, 2001; Bengson *et al*. 2009).

On the one hand, it might seem that one can know how to ride a bike without knowing that something is the case simply by being able to ride a bike, whether or not one knows how one does it or how anyone else does it.

On the one hand, knowing how seems to be like knowing what, knowing where, knowing who, knowing when, and so on, and it would seem that to know these things is to know what the answer is to a corresponding question: "who did it?" "Bob"; "where did he do it?" "Over there"; and "how did he do it?" "Like this." So to know how to ride a bike is to know that this is how to ride a bike. And someone might, in this sense, know how to ride a bike without being able to ride a bike because of lack of practice, fear of falling, or paralysis.

Without trying to resolve this issue, perhaps we can nevertheless at least distinguish *learning how to* ride a bike from *learning to* ride a bike. Albert may have learned how to ride a bike even though he has not learned to ride a bike.

But what sort of learning are we concerned with in this book? We are at least concerned with *learning to do* something and with learning to classify or estimate values. We are concerned with learning from labeled examples to do one or the other of these things.

This sort of learning does not involve learning what the correct classifications of particular examples are. The relations between features and correct classifications are typically only probabilistic, so even a best rule of classification—a Bayes rule—will sometimes, perhaps often, be in error. Furthermore, as we will see, even after seeing many examples, we may not be able to come up with a best rule. So, we cannot in general learn that we have found the Bayes rule or even that we are on the way to finding it.

6.8 QUESTIONS

1. What is the general learning problem we are concerned with?

2. Why do we need learning? Why cannot we simply use a Bayes rule for pattern classification?

3. What are training data for learning from examples?

4. What sorts of assumptions do we need to make about the training data?

5. What does it mean for examples to be drawn to be iid?

6. When might a brute force approach to the learning problem be useful?

7. What is the curse of dimensionality?

8. What is inductive bias? Is it good or bad?

9. Consider a pattern recognition problem with prior probabilities $P(0) = 0.2$ and $P(1) = 0.8$ and conditional distributions $P(x|0)$ and $P(x|1)$. Let R^* denote the Bayes error rate for this problem. As usual, suppose we have independent and identically distributed training data $(x_1, \ell_1), \ldots, (x_n, \ell_n)$. Let M be some universally consistent decision rule for this problem.

   (a) Suppose we randomly select half of the training points and throw them away. Will the remaining training points be independent? Identically distributed?

   (b) If we use method M but on the training data as modified in part (a), what asymptotic error rate will we get?

   (c) Now suppose we take the original training data and throw away all the examples that have a label 1. Will the remaining training data be independent? Identically distributed according to the underlying distributions?

   (d) If we use method M but on the training data as modified in part (c) by throwing away all examples with label 1, what asymptotic error rate will we get?

   (e) Give a very brief explanation of why we get the results of parts (b) and (d).

6.9 REFERENCES

The formulation of learning from examples (or supervised learning) presented here is very standard and is the setting presented in the various references on statistical pattern recognition presented at the end of Chapter 4, some of which are included below. There has been a substantial amount of work done on learning and classification with non-i.i.d. data. Generally the results for dependent data focus on specific problems or methods, and are fairly advanced mathematically. Some of the many pointers to this work can be found in Györfi *et al.* (1989); Holst and Irle (2001); Vidyasagar (2005); Lozano *et al.* (2006); Steinwart *et al.* (2009), and the references therein. There are many papers and books on density estimation. Silverman (1986) and Scott (1992) and references therein provide some pointers to this area. Stanley and Williamson (2001) and Bengson *et al.* (2009) discuss some of the issues described in the appendix.

Bengson J, Moffett MA, Wright JC. The folk on knowing how. Philos Stud 2009;142: 387–401.

Devroye L, Györfi L, Lugosi G. A probabilistic theory of pattern recognition. New York: Springer Verlag; 1996.

Duda RO, Hart PE, Stork DG. Pattern classification. 2nd ed. New York: Wiley; 2001.

Fukunaga K. Introduction to statistical pattern recognition. 2nd ed. San Diego (CA): Academic Press; 1990.

Györfi L, Härdle W, Sarda P, Vieu P. Nonparametric curve estimation from time series, Lecture notes in statistics. Berlin: Springer-Verlag; 1989.

Holst M, Irle A. Nearest neighbor classification with dependent training sequences. Ann Stat 2001;29:(5):1424–1442.

Kulkarni SR, Lugosi G, Venkatesh S. Learning pattern classification—A Survey. IEEE Trans Inf Theory 1998;44(6):2178–2206.

Lozano A, Kulkarni SR, Schapire R. Convergence and consistency of regularized boosting algorithms with stationary β-mixing observations. Adv Neural Inf Process Syst 2006;18:819–826.

Scott DW. Multivariate density estimation: theory, practice and visualization. New York: Wiley; 1992.

Silverman BW. Density estimation. London: Chapman and Hall; 1986.

Stanley J, Williamson T. Knowing how. J Philos 2001;98:411–444.

Steinwart I, Hush D, Scovel C. Learning from dependent observations. J Multivar Anal 2009;100:175–194.

Vidyasagar M. Convergence of empirical means with α-mixing input sequences, and an application to PAC learning. Proceedings of the 44th IEEE Conference on Decision and Control, and the European Control Conference; 2005 Seville, Spain. p 560–565.

CHAPTER 7

The Nearest Neighbor Rule

Recall the pattern recognition problem that we have discussed so far. We are dealing with objects that belong to one of the two classes 0 and 1. We observe a feature vector $\overline{x}$ of an object and we would like to decide whether the object belongs to class 0 or class 1. We want a decision rule that minimizes the probability of error.

If the prior probabilities, $P(0)$ and $P(1)$, and the conditional densities $p(\overline{x}|0)$ and $p(\overline{x}|1)$ are known, then we can implement Bayes decision rule. If these are unknown, then we assume that we have some "training data" $(\overline{x}_1, y_1), (\overline{x}_2, y_2), \ldots, (\overline{x}_n, y_n)$ and wish to come up with a good classification rule on the basis of this data. This is the problem of learning from examples.

In Chapter 6, we described a brute-force approach—first estimate the unknown prior probabilities and conditional densities and then substitute these estimates into the equations for Bayes decision rule. This approach does not work well for reasons described in the previous chapter. In this and in the following chapters, we discuss more practical approaches that bypass the need to first estimate the unknown distributions.

7.1 THE NEAREST NEIGHBOR RULE

Perhaps the simplest decision rule one might come up with is to find in the training data the feature vector $\overline{x}_i$ that is closest to $\overline{x}$, and then decide that $\overline{x}$ belongs to the same class as given by the label y_i. This decision rule is called the "nearest neighbor rule" (NN rule) and can be illustrated by a figure (see Figure 7.1).

In Figure 7.1, the feature vectors are two-dimensional, and so each $\overline{x}_i$ can be represented by a point in the plane. Associated with each $\overline{x}_i$ is a region (called the Voronoi region) consisting of all points that are closer to $\overline{x}_i$ than to any other $\overline{x}_j$. That is, the points in the region associated with $\overline{x}_i$ have $\overline{x}_i$ as their nearest neighbor among the set $\overline{x}_1, \ldots, \overline{x}_n$.

An Elementary Introduction to Statistical Learning Theory, First Edition.
Sanjeev Kulkarni and Gilbert Harman.
© 2011 John Wiley & Sons, Inc. Published 2011 by John Wiley & Sons, Inc.

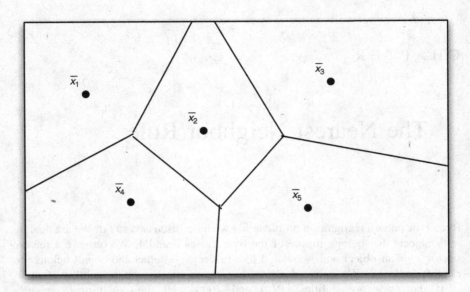

Figure 7.1   Voronoi regions.

Recall that associated with each $\overline{x}_i$ is a label y_i that is either 0 or 1. The NN rule simply classifies a feature vector $\overline{x}$ according to the label associated with the region to which $\overline{x}$ belongs. Alternatively, we can think of the $\overline{x}_i$ as "prototypes." The NN rule classifies a given $\overline{x}$ by assigning it to the same class as the closest prototype.

Given the simplicity of the NN rule, some questions immediately come to mind. For example, is this really a reasonable classification rule? How well does it perform? How should we measure its performance?

7.2   PERFORMANCE OF THE NEAREST NEIGHBOR RULE

As we mentioned above, a natural benchmark for measuring the performance of the NN rule (or any decision rule, for that matter) is the Bayes error rate. Recall that we denote the Bayes error rate by R^*, the "*" indicating optimality (since no rule can have error rate less than R^*).

To discuss the error rate of the NN rule, we need to clarify what we mean. Since the NN rule depends on the data $(\overline{x}_1, y_1), (\overline{x}_2, y_2), \ldots, (\overline{x}_n, y_n)$, the performance of the NN rule will depend on the specific training examples we have seen. Since these training examples are randomly drawn, we cannot expect to do well for every case—there is a chance that we will be unlucky and see particularly poor data. To avoid these problems, we consider the expected performance of the NN rule. The expectation is with respect to the new instance to be classified (as usual) as well as the training examples.

Of course, the expected performance also depends on how much training data have been seen. Let R_n denote the expected error rate of the NN rule after n training examples, and let R_∞ denote the limit of R_n as n tends to infinity. R_∞ is the *asymptotic* expected error rate of the NN rule. That is, it measures the performance of the NN rule as the training sample size tends to infinity. We will focus on R_∞.

What can we say about R_∞? More specifically, can we get bounds on R_∞ in terms of R^*?

A lower bound on R_∞ is easy. Certainly, we must have $R^* \leq R_\infty$, since the Bayes decision rule is optimal and so the nearest neighbor rule (even with infinitely many training examples) can do no better.

The more interesting bounds are upper bounds on R_∞ in terms of R^*. The basic result is that the asymptotic error rate of the NN rule is no worse than twice the Bayes error rate, that is, $R_\infty \leq 2R^*$. Combining the lower and upper bounds, we have

$$R^* \leq R_\infty \leq 2R^*. \tag{7.1}$$

It turns out that one can obtain the following more refined bound:

$$R^* \leq R_\infty \leq 2R^*(1 - R^*). \tag{7.2}$$

Since error rates (and so R^* in particular) must be between 0 and 1, we see that $1 - R^* \leq 1$. Hence, $2R^*(1 - R^*) \leq 2R^*$, so that the simpler bound $2R^*$ follows easily from the more refined upper bound.

The factor of 2 in the upper bound may not seem so good, but for small R^*, even $2R^*$ will be small. Also, in general, we cannot say anything stronger than the bounds given in the sense that there are examples that achieve these bounds. That is, there are choices for the underlying probability distributions for which the performance of the NN rule achieves either the upper or the lower bound.

The upper bounds of Equation (7.1) and (7.2) might seem somewhat surprising. Using only random labeled examples but knowing nothing about the underlying distributions, we can (in the limit) achieve an error rate no worse than twice the error rate that could be achieved by knowing everything about the probability distributions. Moreover, we can do this with an extremely simple rule that bases its decision on just the nearest neighbor to the feature vector we wish to classify. This result is sometimes interpreted informally by saying that half the information is contained in the nearest neighbor.

7.3 INTUITION AND PROOF SKETCH OF PERFORMANCE*

Here we give some intuition and a rough proof sketch on how to obtain the upper bound on the asymptotic error rate.

As we get more and more data, the distance between $\overline{x}$ and its nearest neighbor goes to zero. In computing R_∞, it is as though the distance *is* zero. When the distance between $\overline{x}$ and its nearest neighbor (say $\overline{x}_i$) is zero, it is like guessing the outcome of a coin flip (the label y corresponding to $\overline{x}$) on the basis of another independent coin flip (the label y_i corresponding to $\overline{x}_i$).

Let $R^*(\overline{x})$ be the conditional Bayes error rate, given $\overline{x}$. Then R^* is the average of $R^*(\overline{x})$ over all possible feature vectors $\overline{x}$.

Given $\overline{x}$ (and that the nearest neighbor is distance 0), the probability that the NN rule makes a correct decision is just the probability that both labels are 0 or both labels are 1. That is,

$$P(\text{correct}|\overline{x}) = P(0|\overline{x})P(0|\overline{x}) + P(1|\overline{x})P(1|\overline{x})$$

$$= P(0|\overline{x})^2 + P(1|\overline{x})^2.$$

Now, recall that the Bayes decision rule simply decides the class that has the larger posterior probability, and so makes an error with the probability that is the smaller of the posterior probabilities. Or, conditioned on $\overline{x}$,

$$R^*(\overline{x}) = \min\{P(0|\overline{x}), P(1|\overline{x})\}.$$

Hence, $R^*(\overline{x})$ is equal to one of $P(0|\overline{x})$ or $P(1|\overline{x})$, and $1 - R^*(\overline{x})$ is equal to the other.

In either case, substituting into the expression for $P(\text{correct}|\overline{x})$, we get

$$P(\text{correct}|\overline{x}) = R^*(\overline{x})^2 + (1 - R^*(\overline{x}))^2 \tag{7.3}$$

$$= 1 - 2R^*(\overline{x}) + 2(R^*(\overline{x}))^2. \tag{7.4}$$

Simple Bound
To obtain the simple upper bound that $R_\infty \leq 2R^*$, note that in Equation (7.4) the term $2(R^*(\overline{x}))^2$ is greater than or equal to zero, so that

$$P(\text{correct}|\overline{x}) \geq 1 - 2R^*(\overline{x}).$$

Therefore,
$$P(\text{error}|\overline{x}) = 1 - P(\text{correct}|\overline{x}) \leq 2R^*(\overline{x})$$

and averaging over $\overline{x}$ we get $R_\infty = P(\text{error}) \leq 2R^*$.

Tighter Bound
To get the tighter upper bound of Equation (7.2), we need to be more careful. From Equation (7.4), we get

$$P(\text{error}|\overline{x}) = 1 - P(\text{correct}|\overline{x}) = 2R^*(\overline{x}) - 2(R^*(\overline{x}))^2.$$

Then

$$R_\infty = P(\text{error}) = E[2R^*(\overline{x}) - 2(R^*(\overline{x}))^2]$$
$$= 2R^* - 2E[(R^*(\overline{x}))^2]$$
$$\leq 2R^* - 2(R^*)^2,$$

since the average of squares is greater than or equal to the square of the average. Thus, $R_\infty \leq 2R^*(1 - R^*)$.

7.4 USING MORE NEIGHBORS

Despite its simplicity, the NN rule has impressive performance. Yet, it is natural to ask whether we can do any better. For example, why not use several neighbors, rather than just the *nearest* neighbor?

This is a reasonable suggestion and in fact leads to useful extensions of the (single) NN rule. For example, consider the k-NN rule, in which we use the k nearest neighbors, for some fixed number k. With this rule, given an observed feature vector $\overline{x}$, we use the k nearest neighbors of $\overline{x}$ from among $\overline{x}_1, \ldots, \overline{x}_n$, and take a majority vote of the labels corresponding to these k nearest neighbors. Let R_∞^k be the error rate of the k-NN rule in the limit of infinite data. We might expect that R_∞^k improves (gets smaller) for larger k. This is often the case, but not always. For example, under certain conditions it can be shown that

$$R^* \leq R_\infty^k \leq \left(1 + \frac{1}{k}\right) R^*.$$

However, it can also be shown that there are some distributions for which the 1-NN rule outperforms the k-NN rule for any fixed $k > 1$.

Another very useful idea is to let the number of neighbors used grow with n (the amount of data we have). That is, we can let k be a function of n, so that we use a k_n-NN rule. If we do this, how should we choose k_n?

We need $k_n \to \infty$ so that we use more and more neighbors as the amount of training data increases. But we should make sure that $\frac{k_n}{n} \to 0$, so that asymptotically the number of neighbors we use is a negligible fraction of the total amount of data. This will ensure that we use neighbors that get closer and closer to the observed feature vector $\overline{x}$. For example, we might let $k_n = \sqrt{n}$ to satisfy both conditions.

It turns out that with any such k_n (such that $k_n \to \infty$ and $k_n/n \to 0$ are satisfied), we get $R_n^{k_n} \to R^*$ as $n \to \infty$. That is, in the limit as the amount of training data grows, the performance of the k_n-NN rule approaches that of the optimal Bayes decision rule! What is surprising about this result is that by observing data but without knowing anything about the underlying distributions, asymptotically we can do as well as if we knew the underlying distributions completely. And, this works without assuming that the underlying distributions take on any particular form or satisfy any stringent conditions. In this sense, the k_n-NN rule is called *universally consistent*, and is truly *nonparametric* learning in that the underlying distributions can be arbitrary and we need no knowledge of their form. It was not known until the early 1970s whether universally consistent rules existed, and it was quite surprising when the k_n-NN rules along with some others were shown to be universally consistent. A number of such decision rules are known today.

However, universal consistency is not the end of the story. This is just an asymptotic property (in the limit of infinite data), and "in the long run, we're all dead" (Keynes). A critical issue is that of convergence rates. Many results on the convergence rates of the NN rule and other rules are known. A fairly generic problem is that for most methods the rate of convergence is very slow in high-dimensional spaces. This is a facet of the so-called "curse of dimensionality" mentioned in Section 6.5. As we discussed, in many real applications the dimension can be extremely large, which bodes ill for many methods. Furthermore, it can be shown that there are no "universal" rates of convergence. That is, for any method, one can always find distributions for which the convergence rate is arbitrarily slow. Thus, as we also mentioned in Section 6.5 in the context of "No Free Lunch Theorems," there is no one method that can universally beat out all other methods. These results make the field continue to be exciting, and makes the design of good learning algorithms and the understanding of their performance an important science and art. In the coming chapters, we discuss some other methods useful in practice, as well as some results on what can be said with a finite amount of training data (see Chapters 11, 12, and 13).

7.5 SUMMARY

In this chapter, we discussed a simple learning method that uses training data to come up with a classification rule. The NN rule classifies a new example in the same class as the nearest feature vector in the training data. Surprisingly, this simple rule has a performance no worse than twice that of the optimal Bayes decision rule. We gave a proof sketch for this result as well as slightly more refined result. We then discussed the possibility of using more neighbors. By letting the number of neighbors k_n that we use grow with the data (so that $k_n \to \infty$), but such that $k_n/n \to 0$, we can actually do as well as Bayes decision rule in the limit as $n \to \infty$. No knowledge or assumptions about the underlying distributions are needed, so that this k_n NN rule is an example of what are called universally consistent rules.

7.6 APPENDIX: WHEN PEOPLE USE NEAREST NEIGHBOR REASONING

7.6.1 Who Is a Bachelor?

How do we decide whether someone is a bachelor? One theory is that we have learned a definition or rule: a bachelor is an unmarried adult male person.

But what about the Pope? He is an unmarried adult male person, but is he a bachelor? Many people would say he is not. What about an unmarried adult male who has been living with a woman "as man and wife," although they are not legally married. Many people would not consider that man a bachelor.

Furthermore, what about a man who is married but in the process of getting a divorce, who has not lived with his legal wife for several years and has been dating regularly. Many people say that he is a bachelor even though he is still legally married.

Do people use nearest neighbor reasoning to decide what they think about these cases?

7.6.2 Legal Reasoning

A court wants to decide whether a particular case falls under a legal rule, for example the rule, "No vehicles are allowed in Princeton Park." Some cases are easy to classify, as when someone drives the family sedan into the park. That is clearly forbidden by the rule, but other cases are less clear. Suppose a teenager rides her bicycle into the park. Do bicycles count as *vehicles* under the rule? Do wheelchairs? Rollerblades?

Or consider the legal rule that, if a will clearly specifies that the estate is left to a particular person and that person is living and competent, then the person in question is to inherit the estate. In most cases it is quite clear whether this rule applies. But suppose that the person thus specified in the will has murdered the deceased? Must that person inherit under this rule? Courts have ruled that the person may not inherit.

Or consider the rule that, if one person's wrongful act causes damage to another, then the first person must reimburse the victim for the damage, except to the extent that the victim was responsible for the damage. How does this rule apply to a case in which a driver causes an accident that kills a woman's husband and she has a heart attack on learning of her husband's death? Under this rule must the driver reimburse the wife for her medical expenses and other damage she suffers because of her heart attack?

Various factors enter into legal decisions about such hard cases, especially including *precedent*. A court tries to give the fairest decision, given the facts of the case and given prior decisions that courts have made in similar cases. Is the present case more similar to previously decided case 1, in which case the plaintiff should prevail, or is it more similar to previously decided case 2, in which case the defendant should prevail? Reasoning from precedent is a version of a nearest neighbor strategy for deciding new cases.

7.6.3 Moral Reasoning

In moral reasoning we try to find general principles that apply to particular cases and we also test our general principles against our settled judgments about other cases. So here too is a use of a nearest neighbor strategy.

"Is it morally OK to raise animals for food?" Jack asks Jill. "Of course," she replies. "But would it be morally OK to raise people for food?" "Of course not," she says. Then what is the difference? Perhaps Jill thinks that people are different from (other) animals in morally significant ways. Perhaps it is because people are rational in ways that animals are not rational. But why does that matter? And what about chimpanzees, which seem pretty smart, smarter than young human children? And would it be morally okay to raise brain damaged people for food as long as they will not ever be rational?

In moral reflection of this sort, you try to adjust some of your views in the light of other "nearby" opinions you hold. Perhaps you try to reach a *reflective equilibrium* (Rawls, 1971) between your principles and your particular views about examples so that there is no conflict between principles.

7.7 QUESTIONS

1. (a) What is the NN rule and how does the expected error from the use of this rule compare with the Bayes error rate?
   (b) What conditions on k_n are required for the k_n-NN rule to have an asymptotic error rate equal to the Bayes error rate?

2. Recall that for the 1-NN rule, the region associated with a feature vector $\bar{x}_i$ is the set of all points that are closer to x_i than to any of the other feature vectors $\bar{x}_j$ for $j \neq i$. These are the Voronoi regions. Sketch the Voronoi regions for feature vectors $\bar{x}_1 = (0, 0)$, $\bar{x}_2 = (0, 2)$, $\bar{x}_3 = (2, 0)$, and $\bar{x}_4 = (1, 1)$.

3. Come up with a case (i.e., give the prior probabilities and conditional densities) in which the error rate of the NN rule equals the Bayes error rate, and briefly explain why this happens in the case you give.

4. Describe as precisely as you can the tradeoffs of having a small k_n versus a large k_n in the k_n-nearest neighbor classifier. What happens in the extreme cases when $k_n = 1$ and when $k_n = n$?

5. What conditions are required on k_n for the k_n-NN rule to be universally consistent?

6. If we use a k_n-NN rule with $k_n = n$, what would be the resulting error rate in terms of $P(0)$, $P(1)$, $P(\bar{x}|0)$, and $P(\bar{x}|1)$?

7. Briefly discuss the following position. Under appropriate conditions, the k_n-NN rule is universally consistent, so the choice of features does not matter.

7.8   REFERENCES

Nearest neighbor methods were introduced by Fix and Hodges (1951, 1952) in the early 1950s. Cover and Hart (1967) obtained the now classic result that $R_\infty \leq 2R^*$. A great deal of work on nearest neighbor methods has been done since then. Most books or review papers on statistical pattern recognition discuss nearest neighbor methods. For example, see Chapter 4 of Duda *et al*. (2001) and Section 2 of Kulkarni *et al*. (1998).

Devroye *et al*. (1996) has several in-depth chapters on nearest neighbor methods and their performance as well an extensive bibliography with pointers to original research results. Dasarathy (1991) contains a broad, but less recent, collection of work on nearest-neighbor methods. There is also a useful account in Mitchell (1997).

There is a discussion of "bachelor" in Winograd and Flores (1986), p. 112. For legal reasoning, see e.g. Dworkin (1986), Chapters 1–2. Stich (1993) discusses moral reasoning.

Cover TM, Hart PE. Nearest neighbor pattern classification. IEEE Trans Inf Theory 1967;13(1):21–27.

Dasarathy BV, editor. Nearest neighbor (NN) norms: NN pattern classification techniques. Washington (DC): IEEE Computer Society; 1991.

Devroye L, Györfi L, Lugosi G. A probabilistic theory of pattern recognition. New York: Springer Verlag; 1996.

Duda RO, Hart, PE, Stork, DG. Pattern classification. 2nd ed. New York: Wiley; 2001.

Dworkin R. Law's empire. Cambridge (MA): Harvard UP; 1986.

Fix E, Hodges JL. Discriminatory analysis: nonparametric discrimination: consistency properties. USAF Sch Aviat Med 1951;4:261–279.

Fix E, Hodges JL. Discriminatory analysis: nonparametric discrimination: small sample performance. USAF Sch Aviat Med 1952;11:280–322.

Kulkarni SR, Lugosi G, Venkatesh S. Learning pattern classification—A survey. IEEE Trans Inf Theory 1998;44(6):2178–2206.

Mitchell TM. Instance-based learning. Machine learning. Boston (MA): McGraw-Hill; 1997. pp. 226–229, Chapter 8.

Rawls J. A theory of justice. Cambridge (MA): Harvard UP; 1971.

Stich SP. Moral philosophy and mental representation. In: Hechter M, Nadel L, Michod R, editors. The origin of values. Hawthorne (NY): Aldine de Gruyter; 1993. pp. 215–228.

Winograd T, Flores F. Understanding computers and cognition. Norwood (NJ): Ablex; 1986.

CHAPTER 8

Kernel Rules

In the previous chapter, we discussed the nearest neighbor rule, which is a conceptually simple rule to learn from examples. Nearest neighbor rules fix a number of neighbors and take a majority vote among these neighbors, regardless of how far away they might be from the example to be classified. In this chapter, we start with a variant of this in which we fix a distance and take a majority vote among all neighbors within this distance, regardless of how many there are. This is a special case of a broad and important class of methods called kernel rules.

8.1 MOTIVATION

Recall that in order to classify a feature vector $\bar{x}$, nearest neighbor rules fix a number of neighbors k_n, and then take a majority vote of the labels associated with those k_n feature vectors among the $\bar{x}_1, \ldots, \bar{x}_n$ that are closest to $\bar{x}$. By choosing k_n appropriately (namely, $k_n \to \infty$ and $k_n/n \to 0$), we can guarantee universal consistency. That is, by choosing k_n to satisfy these conditions, we guarantee that, as the amount of training data grows, the performance of the nearest neighbor rule approaches the performance of the optimal Bayes decision rule without any restrictions or prior knowledge regarding the underlying distributions.

These two conditions on k_n ensure that when classifying a feature vector $\bar{x}$, as the data increases we use more and more of the training examples in the majority vote, and yet these examples will become closer and closer to $\bar{x}$. The fact that we use an increasing number of examples is directly enforced by the condition $k_n \to \infty$. The same number of neighbors (namely, k_n) is used regardless of where the feature vector $\bar{x}$ is located.

The fact that the neighbors we use get increasingly close to $\bar{x}$ may not be immediately clear. How far away the k_n neighbors are depends on the location of $\bar{x}$ and on the locations $\bar{x}_1, \ldots, \bar{x}_n$ of the training examples. If $\bar{x}$ is in a region

An Elementary Introduction to Statistical Learning Theory, First Edition.
Sanjeev Kulkarni and Gilbert Harman.
© 2011 John Wiley & Sons, Inc. Published 2011 by John Wiley & Sons, Inc.

where there are very few previous $\overline{x}_i$, then the k_n-th nearest neighbor might not be very nearby at all. Recall that $\overline{x}$ is drawn from some (unknown) probability distribution $P(\overline{x})$ (or density $p(\overline{x})$) and so will be in places where this distribution has nonzero probability (or nonzero probability density in a neighborhood of $\overline{x}$). But the previous examples $\overline{x}_1, \ldots, \overline{x}_n$ are drawn from the *same* distribution. Hence, it is also likely that some of the $\overline{x}_i$ will fall in the same area as $\overline{x}$.

In fact, if a small neighborhood around $\overline{x}$ has some probability α, then we would expect that a proportion roughly α of the n examples (i.e., αn examples) would fall in that small neighborhood. The condition $k_n/n \to 0$ guarantees that the *fraction* of all the examples we consider as neighbors goes to zero since k_n/n is precisely the fraction of training examples that are considered. Alternatively, note that since $k_n/n \to 0$ eventually we have $k_n/n < \alpha$, so that $k_n < \alpha n$. So the number k_n of neighbors used is eventually less than the number αn we expect in the small neighborhood around $\overline{x}$. This argument can be made precise to show that the condition $k_n/n \to 0$ guarantees that with probability approaching 1, the neighbors we use get closer and closer to $\overline{x}$. Of course, exactly how close they are will depend on the locations of $\overline{x}$ and $\overline{x}_1, \ldots, \overline{x}_n$.

8.2 A VARIATION ON NEAREST NEIGHBOR RULES

The above discussion suggests considering a rule slightly different from the nearest neighbor rule that still guarantees that we use an increasing number of examples that get closer to $\overline{x}$. Rather than fixing the number of neighbors, fix a distance h and consider all examples from among $\overline{x}_1, \ldots, \overline{x}_n$ that fall within a distance h of $\overline{x}$. Let us classify $\overline{x}$ according to the majority vote of the labels y_i of all these neighbors within a distance h of $\overline{x}$. If none of the $\overline{x}_i$ falls within a distance h or if there is a tie in the majority vote, we need some way to decide in these cases. Some natural choices are to decide randomly or always select class 0 (or class 1) in such a case.

To specify the rule more precisely, we need some notation. Let $B(\overline{x}, h)$ denote the closed ball of radius h centered at $\overline{x}$. That is, $B(\overline{x}, h)$ contains all the feature vectors that are a distance h or less from $\overline{x}$. Mathematically, given a d-dimensional feature space,

$$B(\overline{x}, h) = \left\{ \overline{z} \in \mathbf{R}^d \mid \|\overline{x} - \overline{z}\| \le h \right\}.$$

Let I_A denote the indicator function of an event A, where $I_A = 1$ if A is true and $I_A = 0$ if A is false. Let

$$v_n^0(\overline{x}) = \sum_{i=1}^n I_{\{y_i=0 \text{ and } \overline{x}_i \in B(\overline{x},h)\}}$$

and let

$$v_n^1(\overline{x}) = \sum_{i=1}^n I_{\{y_i=1 \text{ and } \overline{x}_i \in B(\overline{x},h)\}}.$$

The quantities $v_n^0(\overline{x})$ and $v_n^1(\overline{x})$ denote the vote counts for class 0 and class 1, respectively, that is, the number of examples labeled 0 and 1, respectively, that are within a distance h of $\overline{x}$.

Consider the following classification rule.

$$g_n(\overline{x}) = \begin{cases} 0 & \text{if } v_n^0(\overline{x}) \geq v_n^1(\overline{x}) \\ 1 & \text{otherwise.} \end{cases}$$

In other words, the classification rule $g_n(\overline{x})$ assigns $\overline{x}$ to class 1 if and only if, among the training points within a distance h of $\overline{x}$, there are more points labeled 1 than those labeled 0.

As with nearest neighbor rules, this rule classifies a feature vector $\overline{x} \in \mathbf{R}^d$ according to a majority vote among the labels of the training points $\overline{x}_i$ in the vicinity of $\overline{x}$. However, while the nearest neighbor rule classifies $\overline{x}$ on the basis of a specified number k_n of training examples that are closest to $\overline{x}$, this rule considers all $\overline{x}_i$'s that are within a fixed distance h of $\overline{x}$. The rule we have been discussing is the simplest example of a very general class of rules call kernel rules.

8.3 KERNEL RULES

The simple rule discussed above is sometimes called the *moving window* classifier. Consider a "window function" or "kernel" defined as follows:

$$K(\overline{x}) = \begin{cases} 1 & \text{if } \|\overline{x}\| \leq 1 \\ 0 & \text{otherwise.} \end{cases} \tag{8.1}$$

This function is 1 inside the closed ball of radius 1 centered at the origin, and 0 otherwise (see Figure 8.1). Therefore, the function $K(\overline{x}/h)$ is 1 inside the closed ball of radius h centered at the origin, and 0 otherwise. Then for a fixed $\overline{x}_i$, the function $K(\frac{\overline{z}-\overline{x}_i}{h})$ is 1 inside the closed ball of radius h centered at $\overline{x}_i$, and 0 otherwise (see Figure 8.2). Thus, by choosing the argument of $K(\cdot)$ appropriately, we have "moved" the window to be centered at the point $\overline{x}_i$ and to be of radius h.

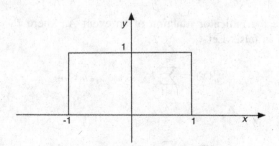

Figure 8.1 Basic window kernel.

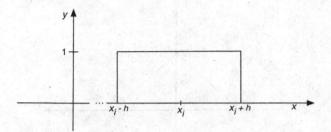

Figure 8.2 Moving window kernels.

We can then write the vote counts $v_n^0(\overline{x})$ and $v_n^1(\overline{x})$ as follows:

$$v_n^0(\overline{x}) = \sum_{i=1}^{n} I_{\{y_i=0\}} K\left(\frac{\overline{x} - \overline{x}_i}{h}\right)$$

and

$$v_n^1(\overline{x}) = \sum_{i=1}^{n} I_{\{y_i=1\}} K\left(\frac{\overline{x} - \overline{x}_i}{h}\right).$$

These are really just different expressions for counting up the votes for the two classes in terms of the kernel function $K(\cdot)$. However, written in this way, it naturally suggests that we might pick a different function $K(\cdot)$. For example, it makes sense that training examples very close to $\overline{x}$ should have more influence (or a larger weight) in determining the classification of $\overline{x}$ than those that are farther away. The moving window rule gives equal weight to all points within a distance h and zero weight to all other points. A smoother transition can be accomplished with different choices for the function $K(\cdot)$. With more general functions, we usually refer to $K(\cdot)$ as a kernel function (or, simply, kernel). We can consider a general kernel function $K:\mathbf{R}^d \rightarrow \mathbf{R}$. Such a function is usually non-negative and is often monotonically decreasing along rays starting from the origin.

The special choice $K(\overline{x}) = I_{\{\overline{x} \in B(0,1)\}}$ is just the moving window rule we started with. Some other popular choices for kernel functions include (see Figures 8.3–8.6)

Triangular kernel: $\quad K(\overline{x}) = (1 - \|\overline{x}\|) I_{\{\|\overline{x}\| \leq 1\}}.$

Gaussian kernel: $\quad K(\overline{x}) = e^{-\|\overline{x}\|^2}.$

Cauchy kernel: $\quad K(\overline{x}) = 1/(1 + \|\overline{x}\|^{d+1}).$

Epanechnikov kernel: $\quad K(\overline{x}) = (1 - \|\overline{x}\|^2) I_{\{\|\overline{x}\| \leq 1\}}.$

The positive number h in $K(\frac{\overline{z} - \overline{x}_i}{h})$ plays a role similar to that in the original case and is called the *smoothing factor*, or *bandwidth*. Recall that in the special case of a uniform kernel, h denoted the distance within which training examples were used

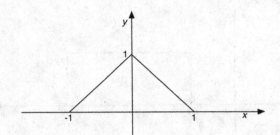

Figure 8.3   Triangular kernel.

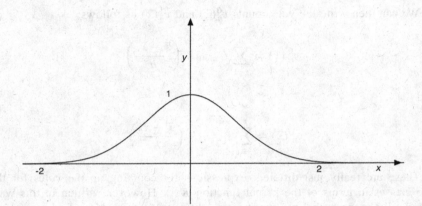

Figure 8.4   Gaussian kernel.

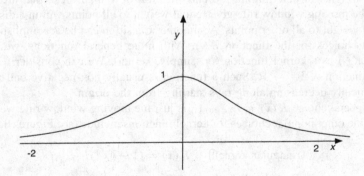

Figure 8.5   Cauchy kernel for d = 2.

in the voting for the classification of $\overline{x}$. It plays a similar role now, but the voting is weighted. The smoothing factor is the most important parameter of a kernel rule. If h is small, the rule gives large relative weight to points near $\overline{x}$, and the decision is very "local," while for a large h many more points are considered with fairly large weight, but these points can be farther from $\overline{x}$. Hence, h determines the amount of

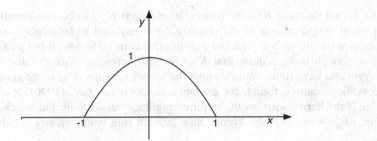

Figure 8.6 Epanechnikov kernel.

"smoothing." In choosing a value for h, one confronts a similar kind of tradeoff as in choosing the number of neighbors using a nearest neighbor rule.

The kernel rule adds up the contributions from the kernel function due to all points labeled 0 and those points labeled 1, respectively, giving the (weighted) votes $v_n^0(\overline{x})$ and $v_n^1(\overline{x})$. The classification of $\overline{x}$ is based on whichever of these is greater.

8.4 UNIVERSAL CONSISTENCY OF KERNEL RULES

As with the nearest neighbor rule, we would like some results on the performance of kernel rules. For example, a natural question is whether there are choices for the kernel function K and the smoothing factor h that lead to universally consistent rules. That is, can we guarantee that as we get more and more training examples, the performance of the kernel rule will approach that of the optimal Bayes decision rule regardless of the underlying probability distributions?

Recall that the smoothing factor h is analogous to the number of neighbors used in nearest neighbor rules. As we might expect, to get universal consistency, we need to let the smoothing factor depend on the amount of data, so we let $h = h_n$.

To make sure that we get "locality" (i.e., so that the training examples used get closer to $\overline{x}$), we need to have $\lim_{n \to \infty} h_n = 0$. To make sure that the number of training examples used grows, we need to have $\lim_{n \to \infty} nh_n^d = \infty$. Recall that $\overline{x} \in R^d$ so that d is the dimension of the feature space. The intuition for this is as follows. Consider the uniform kernel so that a training example is used if it is within a distance h_n of $\overline{x}$. The volume of a d-dimensional ball of radius h_n centered at $\overline{x}$ is proportional to h_n^d. Since with very high probability, the feature vector $\overline{x}$ will fall where the probability density $p(\overline{x})$ is positive, the probability that a particular $\overline{x}_i$ falls in the ball of radius h_n centered at $\overline{x}$ is approximately proportional to the volume of this ball for sufficiently small h_n, and thus is proportional to h_n^d. Hence, out of n training examples, the expected number that fall within h_n of $\overline{x}$ is proportional to nh_n^d. Since this tends to infinity as $n \to \infty$, the number of training examples used also tends to infinity.

These two conditions ($h_n \to 0$ and $nh_n^d \to \infty$) are analogous to the conditions imposed on k_n to get universal consistency. In addition to these two conditions, to show universal consistency we need certain fairly mild regularity conditions on the kernel function $K(\cdot)$. In particular, we need $K(\cdot)$ to be non-negative, and over a small neighborhood of the origin $K(\cdot)$ is required to be larger than some fixed positive number $b > 0$. The last requirement is more technical but as a special case, it is enough if we require that $K(\cdot)$ be non-increasing with the distance from the origin and have finite volume under the kernel function. The more general technical conditions can be found, for example, in Devroye *et al.* (1996). It can be shown that if the kernel satisfies the required regularity conditions and we choose $h_n \to 0$ and $nh_n^d \to \infty$, then the kernel classification rule is universally consistent.

8.5 POTENTIAL FUNCTIONS

The basic kernel classifier computes a vote count for each class by adding up the contributions from the kernel function at each of the training examples. Then a decision is made on the basis of which vote count is larger.

A slightly different formulation is in terms of what is called a *potential function* or *discriminant function*. In this formulation, instead of the vote counts for the two classes, we think of points labeled 1 as adding to the potential function and points labeled 0 as subtracting from it. That is, we compute the potential function

$$f(\overline{x}) = \sum_{i=1}^{n} (2y_i - 1) K\left(\frac{\overline{x} - \overline{x}_i}{h}\right)$$

and then classify $\overline{x}$ according to whether $f(\overline{x}) > 0$ or $f(\overline{x}) < 0$. Note that the term $(2y_i - 1)$ gives 1 if the label $y_i = 1$ and gives -1 if $y_i = 0$. Thus, it provides the positive or negative contribution accordingly.

The reason for the name potential function is that we can think of the points labeled 1 as having a positive charge and those labeled 0 as having a negative charge. The kernel function is thought of as the "potential field" generated by a charge, with the appropriate sign depending on the label. The set of training examples each contribute to the total potential field. Then the point $\overline{x}$ is classified according to whether the potential at $\overline{x}$ is positive or negative. The potential function is really just a different way of expressing the basic kernel rule. If the potential function is positive at $\overline{x}$, then the vote count for class 1 is greater than the vote count for class 0, while the opposite is true if the potential function is negative at $\overline{x}$.

An important extension to the basic kernel rule allows a similar potential function (or vote count) computation to be done using a new set of points rather than the original training examples. First, some other process uses the training data to compute the new set of points $\overline{z}_1, \ldots, \overline{z}_k$, with associated weights $a_1, \ldots, a_k$ and associated "scales" $h_1, \ldots, h_k$. For example, the training examples might be clustered into groups and one representative z_i is chosen for each cluster. The weight a_i might be the number of original training examples in the cluster. The feature vector $\overline{x}$ is then classified according to the sign of the potential function

$$f(\overline{x}) = \sum_{i=1}^{k} a_i K\left(\frac{\overline{x} - \overline{z}_i}{h_i}\right).$$

The number of these new points k does not need to be the same as the number of original training examples n. In fact, often k is much smaller than n. Doing this with $k \ll n$ can provide significant computational advantages.

Other examples that can be formulated in terms of potential functions include radial basis function classifiers (closely related to neural networks) and support vector machines, which we discuss in later chapters.

Such methods have been widely studied for both classification and estimation. Of course, we can obtain the standard kernel classifier as a special case by taking $k = n$, $a_i = 2y_i - 1$, $h_i = h$, and $z_i = x_i$.

8.6 MORE GENERAL KERNELS

We started this chapter by considering a sort of dual of nearest neighbor rules that led to the moving window classifier, and we showed how this could be written in terms of the window function in Equation (8.1). We then discussed how we could use more general kernel functions $K(\overline{x})$. We now describe a further generalization of the form of the kernel function.

So far, all the kernel functions we have considered have been a function of a single variable $\overline{x}$. Moreover, $K(\overline{x})$ was always a function of just $\|\overline{x}\|$, the magnitude of the vector $\overline{x}$. When applied to the data, the argument of the kernel involved $\overline{x} - \overline{x}_i$, so that the computations involved $\|\overline{x} - \overline{x}_i\|$, which is the distance between $\overline{x}$ and $\overline{x}_i$. The shape of typical kernels is such that more weight is given when $\overline{x}$ is close to $\overline{x}_i$, so that $\|\overline{x} - \overline{x}_i\|$ is small. The amount of weight decays (and eventually is zero) as the distance between $\overline{x}$ and $\overline{x}_i$ gets large. However, because the form of the kernels we have been considering involves only $\|\overline{x} - \overline{x}_i\|$, the weight given does not depend on what direction $\overline{x}$ is from $\overline{x}_i$, but depends on only the distance.

It turns out that even more general kernels can be considered that are of the form $K(\overline{x}, \overline{x}_i)$. This allows more fine tuning of how much weight is given depending on $\overline{x}$ and $\overline{x}_i$. With this more general kernel, the scale factors are often absorbed directly into the choice of the kernel function. Since the scale factors generally depend on the number of training examples, we can write $K_n(\overline{x}, \overline{x}_i)$ for the choice of the kernel function.

As we described in the previous section, we could use these more general kernels on a new set of points $\overline{z}_1, \ldots, \overline{z}_k$ instead of the original training data, and we can allow for an associated set of weights $a_1, \ldots, a_k$. Note that the weights a_i are generally functions of the training data, so that $a_i = a_i((\overline{x}_1, y_1), \ldots, (\overline{x}_n, y_n))$. Finally, we can also allow the kernel function to depend on the index i, so that the kernel function is $K_{n,i}(\overline{x}, \overline{x}_i)$. As with the a_i, the choice of $K_{n,i}(\overline{x}, \overline{x}_i)$ would typically depend on the training data $(\overline{x}_1, y_1), \ldots, (\overline{x}_n, y_n)$. For example, the choice of the scale factor that is absorbed in $K_{n,i}(\overline{x}, \overline{z}_i)$ might depend on how many of the training examples are in the vicinity of $\overline{z}_i$.

With the above generality, the potential function becomes

$$f(\overline{x}) = \sum_{i=1}^{k} a_i K_{n,i}(\overline{x}, \overline{z}_i)$$

and the decision rule is as before, namely $\bar{x}$ is classified as 0 or 1, depending on whether $f(\bar{x}) < 0$ or $f(\bar{x}) > 0$, respectively.

This general form of kernel classifier results in the broad class of so-called kernel methods, a special case of which are support vector machines, which are discussed in Chapter 17. Of course, as before, to be able to make statements on the performance of the resulting learning rules, appropriate regularity conditions on the kernel functions and the weights are needed.

8.7 SUMMARY

We started this chapter by motivating the simplest kernel method as a variant of the k_n nearest neighbor rule. Rather than fixing the number of neighbors, in the simplest kernel method we fix a distance h, and classify a feature vector $\bar{x}$ on the basis of a majority vote of the labels of those training examples that fall within h of $\bar{x}$. This method can be viewed as follows. We place a kernel or "window function" that is 1 inside a ball of radius h around each example labeled 0 and add up the contributions. We then do the same thing with all the examples labeled 1. We classify the point $\bar{x}$ according to whether the contributions at $\bar{x}$ from the examples labeled 0 are greater than those from the examples labeled 1.

By allowing more general kernel functions $K(\bar{x})$, we obtain the more general kernel classification rule. These kernels allow smoother transition in the weighting given to examples on the basis of their distance from the point to be classified. An important parameter is the smoothing factor or bandwidth h, which plays a role analogous to that of the distance h in the case of the simplest kernel. Generally, the choice of h will depend on the number of training examples n, so that we write $h = h_n$. By choosing $h_n \to 0$, we guarantee that in the limit only training examples increasingly near to the point being classified will be used. By choosing h_n such that $nh_n^d \to \infty$, we guarantee that in the limit we will use an increasing number of training examples. With these two conditions, which are analogous to the conditions on k_n for the nearest neighbor rule, the kernel rule will be universally consistent. That is, the performance of the kernel rule will approach that of the optimal Bayes decision rule as we get more and more data, no matter what the underlying probability distributions are.

The kernel rules can be recast in terms of potential functions (also called discriminant functions). Furthermore, the kernels can be generalized to the form $K(\bar{x}, \bar{x}_i)$. This, together with the idea of using a new set of points $\bar{z}_1, \ldots, \bar{z}_k$ obtained from the original training examples and the associated weights $a_1, \ldots, a_k$ results in a very general form of so-called kernel methods.

8.8 APPENDIX: KERNELS, SIMILARITY, AND FEATURES

The intuition of nearest neighbor methods is that if feature vectors are nearby one another then the classification of these feature vectors should be similar. That is why the best decision for classifying $\overline{x}$ can be determined by the class of the nearest neighbors of $\overline{x}$.

The idea of kernels extends this idea in various ways, from the simple window function to kernels of the form $K(\overline{x})$ to general kernels of the form $K(\overline{x}, \overline{x}_i)$. We can think of $K(\overline{x}, \overline{x}_i)$ as measuring how similar $\overline{x}$ is to $\overline{x}_i$, or perhaps more correctly how much the classification of $\overline{x}_i$ should influence the classification of $\overline{x}$.

With most commonly used kernels, this measure of similarity is closely related to the Euclidean distance between the feature vectors $\overline{x}$ and $\overline{x}_i$, albeit weighted in some way depending on the exact shape of $K(\overline{x}, \overline{x}_i)$. But why should this be the case? Why should nearby feature vectors be classified similarly?

The answer is that we normally pick features that are aligned well with the structure of the problem. If this is not the case, then the intuition of similar features leading to similar classification can fall apart. For example, consider the following two problems.

In Problem 1, we have a binary feature vector $\overline{x} \in \{0, 1\}^d$. That is $\overline{x} = (x_1, \ldots, x_d)$ with $x_1, \ldots, x_d \in \{0, 1\}$. The classification problem is one version of the exclusive-or problem (which we will see in Chapter 9). In this version, $\overline{x}$ is classified as 0 if $x_1 + \cdots + x_d$ is even and $\overline{x}$ is classified as 1 otherwise.

In Problem 2, the classification problem is the same, but instead of using the features $x_1, \ldots, x_d$, we use a new set of features $z_1, \ldots, z_d$ defined as follows. $z_1 = x_1$, $z_2 = x_1 \oplus x_2$, $z_3 = x_1 \oplus x_2 \oplus x_3, \ldots, z_d = x_1 \oplus x_2 \oplus x_3 \cdots \oplus x_d$, where $\oplus$ denotes the exclusive-or operation just mentioned.

In Problem 1, the classification depends critically on *all* the features. A change in any of the x_i changes the classification of $\overline{x}$. On the other hand, in Problem 2 the only feature that matters is z_d. The original feature vector is classified as 0 or 1 according to z_d. All the other z_i are irrelevant. Thus, the z_i capture the structure of the problem in a way radically different than the x_i. However, no information is lost in this transformation.

Different learning methods will perform very differently depending on whether the features x_i or z_i are used. On the other hand, it might be surprising that universal consistency results guarantee that a universally consistent method will eventually learn a good decision rule, even when features are not well aligned with the structure in the problem. Of course, the catch is in the term "eventually." If the features are not appropriate to the classification problem, then learning can require an exceedingly large number of training examples. The choice of features can change an extremely difficult problem to a trivial one. Thus, in any practical applications, feature selection is a very important consideration. While there are a number of theoretical tools available, a good understanding of the application is crucial and feature selection often is much more of an art than a science.

8.9 QUESTIONS

1. For the simplest kernel classification rule, what choices of the smoothing parameter h are analogous to selecting $k_n = 1$ and $k_n = n$, respectively, in the nearest neighbor rule? What happens to the error in these extreme cases?

2. Write the expression for the vote count for class 0, $v_n^0(x)$ in terms of the data $(x_1, \ell_1), \ldots, (x_n, \ell_n)$, indicator functions, and the kernel $K(x)$.

3. For a one-dimensional feature x, sketch the simple kernel function $K(x) = I_{\{|x| \leq 1\}}$.

4. For a one-dimensional feature x, write the equation for the triangular kernel function, $K(x)$.

5. For the kernel function of the previous problem, smoothing factor $h = 0.5$ and $x_i = 3$, sketch $K(\frac{x-x_i}{h})$.

6. What conditions are required on the smoothing parameter h_n for a kernel rule to be universally consistent?

7. True or False: The decision rules arising from using a kernel method with two different kernel functions, $K_1(x)$ and $K_2(x) = 2K_1(x)$, are exactly the same.

8. True or False: There might be some set of labeled data (training examples) such that the 1-nearest neighbor method and the Kernel method (for some $K(x)$ of your choice) can give exactly the same decision for any observed feature vector.

9. Consider the special case where we have a 1-dimensional feature vector and are interested in using a kernel rule. Suppose we have the training data (0,0), (1,1), and (3,0), and we use the simple moving window classifier (i.e., with kernel function $K(x) = 1$ for $|x| \leq 1$ and $K(x) = 0$ otherwise).

   (a) Sketch the functions $v_3^0(x)$ and $v_3^1(x)$ for $h = 0.5$ and the classification rule. Indicate where there are ties. (Note the subscript simply denotes the fact that we have three training examples.)
   (b) Repeat part (a) for $h = 1$.
   (c) Repeat part (a) for $h = 2$.
   (d) What happens for $h \geq 2$?

10. Repeat parts (a), (b), and (c) of the previous problem for the triangular kernel. For part (d), find the value h for which the classification rule first decides 0 for all x.

8.10 REFERENCES

Kernel classification rules have their origin in the kernel density estimates of Parzen (1962), Rosenblatt (1956), and Akaike (1954); in the analogous regression estimators of Nadaraya (1964,1970) and Watson (1964); and in *potential function methods* of Aizerman *et al.* (1964), Bashkirov *et al.* (1964), and Braverman (1965).

There has been a great deal of work on kernel methods for classification and regression, including the classical estimators described here as well as more general kernel-based methods. For extensive bibliographies we refer to Devroye *et al.* (1996) and Kulkarni *et al.* (1998) for classification, Györfi *et al.* (2002) for regression estimation, and Schölkopf and Smola (2001) and Shawe-Taylor and Cristianini (2004) for support vector machines and other kernel-based methods for classification and regression.

Aizerman MA, Braverman EM, Rozonoer LI. Theoretical foundations of the potential function method in pattern recognition learning. Autom Remote Control 1964;25:917–936.

Akaike H. An approximation to the density function. Ann Inst Stat Math 1954;6:127–132.

Bashkirov O, Braverman EM, Muchnik IE. Potential function algorithms for pattern recognition learning machines. Autom Remote Control 1964;25:692–695.

Braverman EM. The method of potential functions. Autom Remote Control 1965;26:2130–2138.

Devroye L, Györfi L, Lugosi G. A probabilistic theory of pattern recognition. New York: Springer-Verlag; 1996.

Györfi L, Kohler M, Krzyżak A, Walk H. A distribution-free theory of nonparametric regression. New York: Springer-Verlag; 2002.

Kulkarni SR, Lugosi G, Venkatesh S. Learning pattern classification—A survey. IEEE Trans Inf Theory 1998;44(6):2178–2206.

Nadaraya EA. On estimating regression. Theory Probab Appl 1964;9:141–142.

Nadaraya EA. Remarks on nonparametric estimates for density functions and regression curves. Theory Probab Appl 1970;15:134–137.

Parzen E. On the estimation of a probability density function and the mode. Ann Math Stat 1962;33:1065–1076.

Rosenblatt M. Remarks on some nonparametric estimates of a density function. Ann Math Stat 1956;27:832–837.

Schölkopf B, Smola AJ. Learning with Kernels: support vector machines, regularization, optimization, and beyond (adaptive computation and machine learning); Cambridge (MA):MIT Press 2001.

Shawe-Taylor J, Cristianini N. Kernel methods for pattern analysis. Cambridge: Cambridge University Press; 2004.

Watson GS. Smooth regression analysis. Sankhya Ser A 1964;26:359–372.

CHAPTER 9

Neural Networks: Perceptrons

Neural networks (or neural nets or artificial neural networks) are collections of (usually simple) processing units, each of which is connected to many other units. The modern study of neural networks is often considered to have begun in 1943 with the publication of a paper by McCulloch and Pitts. Such a network resembles a brain with neurons as processing units and synapses as connections between the units.

One of the original motivations for studying such networks was to model the brain, which is the origin of the term "neural networks" and other terminology in this area. Other motivations include the study of massively parallel and distributed computation, and the use of such networks for learning problems. To distinguish these motivations from that of modeling the brain, sometimes the term "artificial neural networks" is used, although it is simpler and more common to drop the term "artificial" as we do.

9.1 MULTILAYER FEEDFORWARD NETWORKS

A human brain has on the order of 10^{11} (100 billion) neurons and 10^{16} (10 quadrillion) synapses. Neural nets used in practical learning problems have far fewer units and connections. However, even with tens or hundreds of units and several times that many connections, a network rapidly can get complicated.

A small network with a fairly simple structure is shown in Figure 9.1. The processing units (neurons) are depicted by circles. The connections (synapses) are indicated by arrows. These are directed connections, in which the output of one neuron (without the arrow head) serves as an input to another neuron (with the arrow head). Bidirectional connections can be obtained by adding another arrow in the opposite direction. We discuss the details of the inputs and outputs in the next section, but first we discuss some general issues regarding the architecture of the network.

An Elementary Introduction to Statistical Learning Theory, First Edition.
Sanjeev Kulkarni and Gilbert Harman.
© 2011 John Wiley & Sons, Inc. Published 2011 by John Wiley & Sons, Inc.

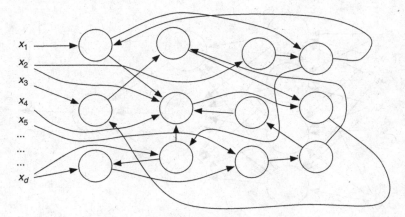

Figure 9.1 Neural net.

Understanding the behavior of even small networks such as that in Figure 9.1 can be difficult. One immediate problem is that there is no clear "output" of the network. This could be easily solved by designating one or more of the neurons as output neurons, and letting the network output be defined by the computation performed by the output units. However, a more fundamental problem is that there are "feedback" loops in the connectivity structure of the network. The output of a neuron can serve as inputs to others, which (via still others) come back and serve as inputs to the original neuron. The existence of such loops makes it difficult to define the "result" of the computation performed by the network. The network is a dynamic system that may or may not settle into some stable result.

In some applications, such dynamic properties can be useful, but we consider a special class of networks that avoids the problems caused by feedback loops—multilayer feedforward networks (Figure 9.2).

The units in a multilayer feedforward network are organized in layers, with the output of neurons in one layer serving as the inputs to the neurons in the next layer. Because there are no feedback loops, the behavior of the network is simple to compute.

The first layer receives external inputs, which, as we are going to see shortly, will be features of the objects we wish to classify. The output of each unit in this layer is computed. Once we know these output values, we know all of the inputs to the units in the next layer. Proceeding in this manner, we eventually find the outputs of the last layer (called the output layer), which gives the final result computed by the network for the given input.

9.2 NEURAL NETWORKS FOR LEARNING AND CLASSIFICATION

Feedforward neural networks can be used to solve the pattern recognition problem discussed earlier. The inputs to the first layer are the features of the object we wish

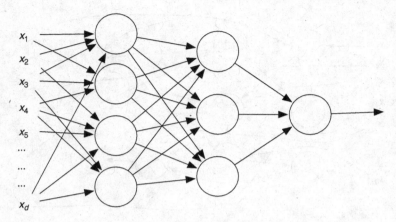

Figure 9.2 Feedforward network.

to classify. The last layer is the output layer, which in our case consists of just one unit. Therefore, there is a single output value for the network, which gives the class to which the network assigns the input feature vector. In some cases, the output value will take on only two values (say 0 and 1, or -1 and 1). In such cases, the classification of the network is obvious. An output of 0 (or -1) means that the input feature vector is assigned to class 0, while an output of 1 means that the feature vector is assigned to class 1.

In some cases, the output of the network might take on arbitrary real values. In this situation, another step called *thresholding* is generally performed to obtain the classification. This involves comparing the output of the network with some value called a threshold. If the output is larger than the threshold, the input feature vector is assigned to class 1; otherwise, it is assigned to class 0.

Given any feature vector, if we apply this feature vector as an input to the network, then we will get one of the two classes as the output of the network. The network implements a classification rule, mapping feature vectors to $\{0, 1\}$.

The particular classification rule implemented by a network is determined by the specifics of the network architecture and the computation done by each neuron. A crucial part of the network computation is a set of parameters called weights. There is usually a real-valued weight associated with each connection between units. The weights are generally considered to be adjustable, while the rest of the network is usually thought of as fixed. Thus, we think of the classification rule implemented by the network as being determined by the weights, and this classification rule can be altered if we change the weights.

To solve a given pattern recognition problem with a neural net, we need a set of weights that results in a good classification rule. As we discussed previously, it is difficult to directly specify good decision rules in most practical situations, and this is made even more difficult by the complexity of the computation performed by the neural net. So, how should we go about selecting the weights of the network?

This is where learning comes in. Suppose we start with some initial set of weights. It is unlikely that the weights we start with will result in a very good classification rule. But, as before, let us assume that we have a set of labeled examples (training data). If we use this data to "train" the network to perform well on this data, then perhaps the network will also "generalize" and provide a decision rule that works well on new data. The success of neural networks in many practical problems is due to an efficient way of training a given network to perform well on a set of training examples, together with the fact that in many cases the resulting decision rule does in fact generalize well.

9.3 PERCEPTRONS

A natural place to start a more detailed discussion is a single unit. A perceptron, depicted in Figure 9.3, is one such unit with simple behavior.

The inputs are denoted by $x_1, x_2, \ldots, x_d$, which, as mentioned before, make up the feature vector $\overline{x} \in \mathbf{R}^d$. Each input x_i is connected to the unit with an associated weight w_i.

The output a of the perceptron is given by

$$a = \text{sign}(x_1 w_1 + \cdots + x_d w_d).$$

The function $\text{sign}(\cdot)$ returns the sign of the input argument. That is,

$$\text{sign}(u) = \begin{cases} -1 & \text{if } u < 0 \\ 1 & \text{otherwise.} \end{cases}$$

In the expression for the output of a perceptron, the argument of the sign function is just a weighted combination of the features. Each feature x_i is multiplied by the corresponding weight w_i and these products are then summed. The output of the perceptron is either -1 (which corresponds to class 0) or 1 (which corresponds to class 1), depending on the sign of the weighted combination $x_1 w_1 + \cdots + x_d w_d$.

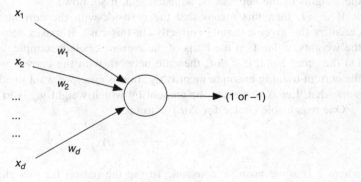

Figure 9.3 Perceptron.

9.3.1 Threshold

The sort of perceptron just described has a threshold of 0. It outputs 1 if the weighted combination of inputs is greater than 0 and outputs -1 if the weighted combination of inputs is less than 0. Perceptrons can have other thresholds as well, for example, 0.5, in which case the output is 1 if the weighted combination of the inputs is greater than or equal to 0.5.

Often it is convenient to consider only units with a threshold of 0, as we do here. One might think that this is rather restrictive. However, if we allow adding another input, then a unit with a nonzero threshold can be mimicked by a unit with a zero threshold (and an extra input). This is done by setting the value of the extra input, say x_0, to the constant value 1. Then the weight w_0 associated with this input effectively serves as a threshold. The output of the perceptron is 1 if

$$x_0 w_0 + x_1 w_1 + \cdots + x_d w_d \geq 0.$$

Since $x_0 = 1$, the output is 1 if

$$x_1 w_1 + \cdots + x_d w_d \geq -w_0.$$

So this behaves just like a perceptron with threshold $-w_0$.

9.4 LEARNING RULE FOR PERCEPTRONS

As discussed, the classification rule computed by the perceptron is stored in (or, more precisely, determined by) the weights. Learning involves adapting the weights on the basis of available training data. Suppose we consider the examples in the training data one by one. Let $\overline{x} = (x_1, \ldots, x_d)$ denote the feature vector of the current example under consideration and let t denote the label corresponding to this example. t stands for "target" and is either -1 or 1, indicating the class to which the given feature vector is assigned. On the basis of this example, should the weights of the network be adjusted and, if so, how?

If $a = t$, then this means that the network with the current weights already classifies the given example correctly. In this case, it makes sense not to change the weights, at least on the basis of the current labeled example.

On the other hand, if $a \neq t$, then the network with the current weights classifies the current training example incorrectly. In this case, we want to adjust the weights somewhat. Let Δw_j denote the amount by which weight w_j is to be adjusted.

One reasonable choice for Δw_j is to take

$$\Delta w_j = c(t - a)x_j. \tag{9.1}$$

where c is some positive constant. To see the reason for this choice, consider a special case. Suppose that the target class is $t = 1$ but that the network computes an output $a = -1$. This means that the weighted sum $x_1 w_1 + \cdots + x_d w_d$ is negative, but to get a correct classification on this example, the weighted sum would need to be positive. (Recall that we are assuming a threshold of 0.)

Therefore, if we are going to adjust the weight w_j, it would make sense to do so in a way that increases the weighted sum $x_1 w_1 + \cdots + x_d w_d$. The only influence w_j has on the weighted sum comes through the term $x_j w_j$. If we change w_j to $w_j + \Delta w_j$, then the term $(\Delta w_j) x_j$ gets added to the weighted sum.

If the feature x_j is positive, we can increase the weighted sum by increasing the weight w_j (i.e., by having Δw_j be positive). But, if x_j is negative, to increase the weighted sum we need to decrease w_j (i.e., have Δw_j be negative). Hence, for $t = 1$ and $a = -1$, the sign of Δw_j should be the same as the sign of x_j. The choice for Δw_j given in Equation (9.1) does exactly this, since c is positive and $t - a = 2$ is also positive.

Now, consider the case that $t = -1$ but $a = 1$. In this case, the weighted sum $x_1 w_1 + \cdots + x_d w_d$ is greater than 0 (since we have the actual output $a = 1$), but we need the weighted sum to be less than 0 (since the target output is $t = -1$). Therefore, we would like to decrease the weighted sum. This can be accomplished by having Δw_j have the opposite sign as x_j, so that the quantity $(\Delta w_j) x_j$ (the amount by which the weighted sum changes) is negative. Again, the choice given above for Δw_j accomplishes this, since in this case the term $t - a = -2$ is negative.

Although the adjustment to the weights given above is in the right direction, it still may not be enough to make the network classify the example correctly. This is actually a good thing, since otherwise the weights might get changed drastically back and forth as we use each successive labeled example. By choosing a constant c that is not too large, we make gradual adjustments to the weights. But to make sure that we do a good job in classifying all the training examples, we cycle through the training examples several times.

To summarize, this learning rule, called the Perceptron Convergence Procedure, is as follows:

Perceptron Convergence Procedure

1. Choose some initial values for all the weights and for the constant c.
2. Repeatedly cycle through the training examples. For each training example considered, adjust the weights according to the equation

$$\Delta w_j = c(t - a)x_j. \tag{9.2}$$

It can be shown that after cycling through the training examples sufficient number of times, this "learning rule" will eventually classify all the training data correctly *assuming that the perceptron is capable of classifying the training data correctly*. When this result was obtained by Rosenblatt 1960, it generated a great deal of excitement over perceptrons, by showing that a simple learning rule was available through which a perceptron would provably learn to classify training data correctly whenever it was possible for a perceptron to classify the data correctly.

9.5 REPRESENTATIONAL CAPABILITIES OF PERCEPTRONS

The excitement over perceptrons was forcefully diminished with the 1969 publication of *Perceptrons* by computer scientists Minsky and Papert. In this book, they observed that the condition "assuming that the perceptron is capable of classifying the training data correctly" is severely limiting. That is, they observed that the class of decision rules that a perceptron is capable of representing (through adjustments of the weights) is very restrictive.

Recall that the output of a perceptron is given by

$$a = \text{sign}(x_1 w_1 + \cdots + x_d w_d).$$

For a fixed set of weights, the set of feature vectors gets partitioned into two classes based on the sign of the weighted sum $x_1 w_1 + \cdots + x_d w_d$. Specifically, the feature vector $(x_1, \ldots, x_d)$ gets classified as follows:

$$x_1 w_1 + \cdots + x_d w_d \geq 0 \implies \text{class 1}$$

and

$$x_1 w_1 + \cdots + x_d w_d < 0 \implies \text{class 0}.$$

The boundary of these two classes is given by the equation

$$x_1 w_1 + \cdots + x_d w_d = 0,$$

which describes a hyperplane in $\mathbf{R}^d$. Thus, no matter how we adjust the weights, the only types of decision regions the perceptron can represent are those with a hyperplane decision boundary. In fact, with a threshold of 0 as we have been assuming, the hyperplane must pass through the origin. In the general case, with a nonzero threshold, the decision boundary must still be a hyperplane but the hyperplane need not go through the origin. However, by adding an extra input as we discussed before (which corresponds to increasing the dimension by one), a hyperplane decision rule in $\mathbf{R}^d$ not passing through the origin can be mimicked by one in $\mathbf{R}^{d+1}$ that does pass through the origin.

In other words, the only time a perceptron can classify all the training examples correctly is when the examples are *linearly separable*. This means that there is a hyperplane in $\mathbf{R}^d$ that separates the training examples such that all the examples labeled $+1$ are on one side of the hyperplane, and those labeled -1 are on the other side of the hyperplane. Figure 9.4 shows this idea in two dimensions, in which case the hyperplane is just a straight line. A perceptron can only represent classes that are linearly separable and so can only "learn" classes that are linearly separable.

One well-known example of a problem that cannot be solved by perceptrons is the so-called exclusive-OR (XOR) problem. Suppose we have just two features x_1 and x_2. Feature vectors belong to class 1 if x_1 or x_2 (but not both) are positive. Then, as depicted in Figure 9.5, the first and third quadrants correspond to class 0, while the second and fourth quadrants correspond to class 1. Clearly, these regions

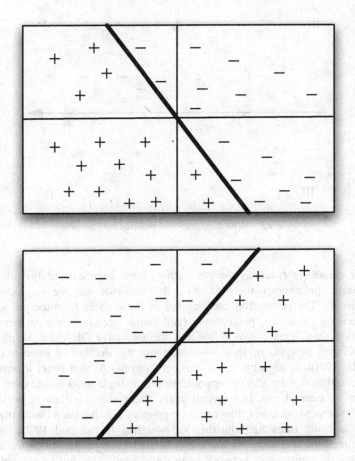

Figure 9.4 Hyperplane decision rules. In $\mathbf{R}^2$ these are just straight lines.

are not linearly separable, and so perceptrons cannot even represent the decision rule needed to solve this simple problem.

Several points need to be made. As discussed in Chapter 4, for various reasons in most practical problems the two classes are not completely disjoint. There is typically overlap, and we use a probabilistic formulation to model the relationships between the feature vectors of objects and the class to which the object belongs. Thus, generally, we should not expect training data to be linearly separable.

In fact, forcing a learning algorithm to perfectly separate the training data is often a bad thing to do, since it tends to "fit to noise" and results in overly complex rules that do not perform well on future (unseen) examples. What is more important than perfect separation of the training data is to come up with good classifiers in the sense of Chapter 4 (e.g., the optimal Bayes decision rule).

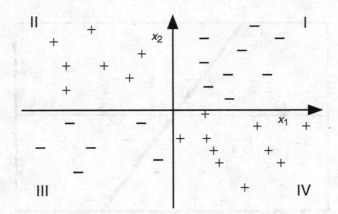

Figure 9.5   XOR Representation problem for perceptrons.

Linear classifiers (such as the perceptron) have been extremely well studied, and in some applications, the best linear decision rule may be good enough (or even optimal). The perceptron can be used in these cases to come up with good linear decision rules. So, perceptrons (and linear decision rules in general) are more useful than linear separability and the exclusive-OR example might lead one to believe. In spite of this, some attribute the decline of interest in neural nets in the 1970s to negative results on perceptrons. At that time, it was unclear whether multilayer networks (as opposed to just a single unit) would overcome the limitations of perceptrons. For several years, there were no clear negative results on multilayer networks, but also no clear positive results, because of the lack of adequate learning rules for adjusting the weights. In the mid 1980s, results in several directions including an algorithm (called back-propagation) for adjusting the weights of a multilayer network were discovered. This led to an explosion of interest in neural nets and is the subject of the next chapter.

9.6   SUMMARY

Some classification functions can be represented as feedforward multilayer neural networks. Given the data about an object to be classified, the network outputs a decision about the classification of the object. Such networks can be "trained" to give good classifications by trying them out on known cases and by adjusting weights of connections in the network so as to improve the resulting classifications on the training data. We started by considering how to train a very simple "network" consisting of just a single unit called a perceptron. This method allows a perceptron to be trained so as to eventually result in a rule that correctly classifies the training examples, whenever it is possible for the perceptron to separate the training examples. Unfortunately, perceptrons are quite limited in the classification rules that they can represent, being able to represent only linear (hyperplanar) decision rules. In the next chapter, we consider multilayer networks that overcome the representational limitations of perceptrons.

9.7   APPENDIX: MODELS OF MIND

Neural net models of mind are different from many earlier models. Here we provide a short account of the differences.

Descartes (1641) made a sharp distinction between mind and body—between the ways in which conscious thoughts could be explained and the ways in which physical events could be explained. Conscious reasoning and choice was to be explained in terms of reasons and free will. Physical events could be explained purely mechanically. Mental and physical events were therefore subject to different sorts of principles. There was an interesting question how the mind and body might interact, as when the mind is affected by perception of physical events and when the body is affected by mental decisions to act in one or another way. Descartes thought that the locus of interaction was the pineal gland in the brain, although the principles of such causal interaction were unclear.

Worries about Cartesian dualism led some theorists to propose that the mind is simply an aspect of the physical brain. Thoughts and sensations were to be identified with physical events. But it was unclear how such an identification could be possible.

The digital computers of the 1940s and 1950s led many theorists to think of the brain as a serial process digital computer and to identify thoughts and other mental events with the sorts of events that occur when a digital computer runs a program. Some models of mind in that period agreed with Descartes that mental processes are conscious processes of thought, especially reasoning, which might be identified with *symbol processing*, in a computer, where the symbols are words and phrases in a natural language or the sorts of symbols used in mathematics and formal logic.

For a while, researchers developed models of what they took to be aspects of mind by writing computer programs that might simulate how people consciously solve certain sorts of problems. One hope at the time was that such programs would constitute substantive psychological theories of how people reason. Another was that the programs would result in a kind of *artificial intelligence* that did better than people at certain tasks.

Soon such models were extended to *production systems* using *parallel processing* by *spreading activation*. Production rules would *fire* when certain conditions were met by activated symbolic expressions; the results of such firing would make more symbolic expressions active; this sort of process would amount to a kind of inference or reasoning.

In these models not all mental activity was conscious. Various expressions might be activated for processing without becoming conscious. Various theories were put forward as to which of these representations should be counted as conscious.

These were all models based on treating the mind as a serial digital computer and they all supposed that thinking involved the processing of symbolic representations that are similar to linguistic or logical or mathematical representations.

Neural net models differ in that they are based on a certain simplified conception of the way the brain works using nonsymbolic processing. In the symbolic processing models derived from thinking of the brain as something like a computer, the knowledge in a system is explicitly represented as symbols in memory or as rules that operate on symbols. But neural net models take the relevant processing to be nonsymbolic. There are many connections among neurons allowing activation in some neurons to activate others to a greater or lesser degree, possibly inhibiting activation in other neurons.

The information in an artificial neural network is not contained in symbols that are stored somewhere or activated. The information is implicitly represented in the weights of connections among nodes in the network. The information is distributed among the weights of connections between neurons. Particular weights do not symbolize anything.

Sometimes a neural network will activate processes that amount to conscious thinking of certain thoughts that may contain symbolic representations. But most of the processes in a neural network will not be like that.

There is still an issue about how activities in a neural network could give rise to or be the same as conscious thinking. But we will not try to resolve that issue.

9.8 QUESTIONS

1. (a) Sketch a diagram of a perceptron with three inputs x_1, x_2, x_3 and weights w_1, w_2, w_3. Label the inputs, weights, and output.
   (b) Write the expression for the output in terms of the inputs and the weights, assuming a threshold of 0.
   (c) What is the output when the inputs are $-3, 2, 1$ and the weights are $0, 0.2, -0.2$?
   (d) What is the output if w_1 is increased to 0.1, but everything else is kept the same?

2. Consider a linear threshold unit (perceptron) with three inputs and one output. The weights on the inputs are respectively 1, 2, and 3, and the threshold is 0.5. If the inputs are respectively 0.1, 0.2, and 0.3, what is the output? If all the three inputs are 0.1, what is the output?

3. Design a linear threshold unit with two inputs that outputs the value 1 if and only if the first input has a greater value than the second. (What are the weights on the inputs and what is the threshold?)

4. Consider a classification problem in which each instance consists of d features $x_1, \ldots, x_d$, each of which can take on only the values 0 or 1. Come up with a

linear threshold unit (a single perceptron) that functions as an AND gate (i.e., the output is 1 if all of the inputs x_i are 1, and the output is zero otherwise). Repeat for an OR gate (i.e., the output is 1 if one or more of the inputs x_i are 1, and the output is zero only if all the inputs are 0).

5. As in the previous problem, consider a classification problem in which each instance consists of d features $x_1, \ldots, x_d$, each of which can take on only the values 0 or 1. A feature vector belongs to class 0 if $x_1 + x_2 + \cdots + x_d$ is even (i.e., the number of 1's is even) and it belongs to class 1 otherwise. Can this problem be solved by a single perceptron? A three-layered network? Why or why not?

6. Consider the following network. There are four inputs with real values. Each input is connected to each of two perceptrons on the first layer that do not take thresholds but simply output the sum of the weighted products of their inputs. Each of these perceptrons is connected to an output threshold unit. Show that this network is equivalent to a network with a single unit, which is a threshold perceptron unit.

7. How is a perceptron trained? What features of the perceptron typically change during training? Formulate a learning rule for making such changes and explain how it works.

9.9 REFERENCES

The paper by McCulloch and Pitts (1943) is often considered to have spawned the modern study of neural networks. In several papers published around 1960, Rosenblatt (1958, 1960, 1962) published results on the perceptron. Other early related results were obtained around the same time by Widrow and Hoff (1960), Widrow (1962). The original publication of the book by Minsky and Papert was in 1969, with the expanded edition listed below appearing in Minsky and Papert (1988). Since the mid 1980s, there has been an explosion of work on neural networks, including several journals and conferences devoted to this subject. There are many books on the subject. Russell and Norvig (2010), Mitchell (1997) and Duda *et al.* (2001) provide readable treatments. A comprehensive treatment is provided by Haykin (1994).

Descartes R. Meditationes de Prima philosophia. Paris; 1641.

Duda RO, Hart PE, Stork DG. Pattern classification. 2nd ed. New York: Wiley; 2001.

Haykin S. Neural networks: a comprehensive foundation. New York: Macmillan Publishing Company; 1994.

McCulloch WS, Pitts W. A logical calculus of the ideas immanent in nervous activity. Bull Math Biophys 1943;5:115–133.

Minsky M, Papert S. Perceptrons. Expanded edition. Cambridge, MA: MIT Press; 1988.

Mitchell T. Machine learning, Boston (MA): McGraw-Hill; 1997. pp. 226–229.

Rosenblatt F. The perceptron: a probabilistic model for information storage and organization in the brain. Psychol Rev 1958;65:386–408.

Rosenblatt F. Perceptron simulation experiments. Proc Inst Radio Eng 1960;48:301–309.

Rosenblatt F. Principles of neurodynamics. Washington, DC: Spartan Books; 1962.

Russell S, Norvig P. Artificial intelligence: a modern approach, Chapter 18. Upper Saddle River, NJ: Prentice-Hall; 2010, pp. 645–767.

Widrow B, Hoff ME. Adaptive switching circuits. IRE WESCON Convention Record 1960;4:96–104.

Widrow B. (1962) Generalization and information storage in networks of adaline 'neurons'. In: Yovitz MC, Jacobi GT, Goldstein GD (eds). Self-organizing Systems. Washington, DC: Spartan; pp. 435–461.

CHAPTER 10

Multilayer Networks

In the previous chapter, we saw that a perceptron can form only linear decision rules. Although such decision regions can sometimes be useful, there are a number of applications in which the restriction to linearly separable classes is very limiting.

One way to overcome the limitation of linearly separable rules is to consider networks consisting of multiple interconnected perceptrons. As discussed in Chapter 9, the special class of multilayer feedforward networks is particularly useful. An example of such a network is shown in Figure 10.1.

Each node of the network behaves as a perceptron. That is, each input is multiplied by the corresponding weight, and the output of the unit is either -1 or 1, depending on whether the weighted sum of the inputs is less than a threshold, for example, 0, or greater than the threshold. The output of the network can be easily computed once the inputs and weights are specified.

10.1 REPRESENTATION CAPABILITIES OF MULTILAYER NETWORKS

Since one of our motivations for considering multilayer networks is to avoid the perceptron's limitation to linear decision regions, the first question that arises is "What kinds of decision regions can a multilayer network represent?"

Perhaps surprisingly, the answer is that with just three layers and enough units in each layer, a multilayer network can approximate any decision rule. In fact, the third layer (the output layer) consists of just a single output unit, so multiple units are needed only in the first and second layer. For decision rules that assign to class 1 a possibly infinite number of areas in the plane (or volumes in N-space), this result can be shown relatively simply as follows.

First, we can approximate any decision rule that assigns feature vectors in some convex set to class 1 and all others to class 0. (A set of points is convex if a

An Elementary Introduction to Statistical Learning Theory, First Edition.
Sanjeev Kulkarni and Gilbert Harman.
© 2011 John Wiley & Sons, Inc. Published 2011 by John Wiley & Sons, Inc.

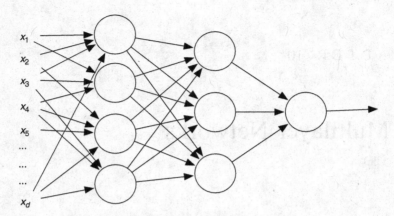

Figure 10.1   Feedforward network with input $\overline{x} = (x_1, \ldots, x_d)$.

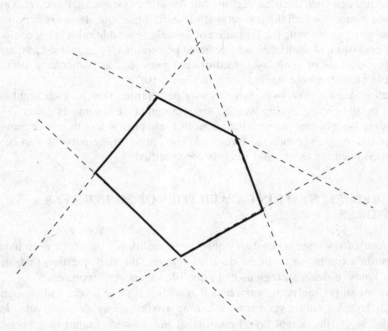

Figure 10.2   Creating a polyhedron by intersecting half-spaces.

line between any two points in the set is always entirely within the set.) This is because a convex set can be approximated by a polyhedron obtained by intersecting a number of half-spaces (Figures 10.2 and 10.3). Each half-space has a linear decision boundary that can be implemented by a single perceptron, as discussed in the previous chapter. The various half-spaces used to approximate the convex set are implemented by different units all in the first layer.

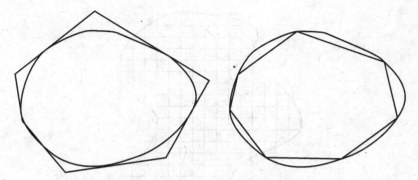

Figure 10.3 Two common ways to approximate a convex set by a polyhedron.

To get the intersection of the half-planes, we pass the outputs of the perceptrons in the first layer to a single perceptron in the second layer that computes the logical AND of its inputs, so that the output is 1 if and only if every input to the single perceptron in the second layer is 1. This happens exactly when the feature vector lies inside the polyhedron defined by the half-spaces. By taking more units in the first layer, we get a polyhedron with more faces, and therefore, a better approximation to the convex set. (The problem of how to make a perceptron compute the AND function was an exercise in the preceding chapter.)

To approximate an arbitrary (nonconvex) set, we can first approximate it with a union of convex sets (Figure 10.4). One systematic way to do this is to partition the space regularly along each dimension to form hypercubes. Each of the constituent convex sets can be approximated as described above, so we only need a way to take the union of these sets. To get the union, we can use a single perceptron that computes the logical OR of its inputs. That is, the output of such a perceptron is 1 if any input is 1. (The problem of how to make a perceptron compute the OR function was also an exercise in the preceding chapter.)

Finally, the network used to approximate a general set has the following form. Each unit in the first layer computes a half-space. The outputs of this layer are passed to the second layer, where each unit performs the logical AND of those half-spaces needed to approximate various convex sets. The outputs from the second layer are then passed through a final unit that performs an OR operation in order to take the union of those convex sets. See Figure 10.5.

10.2 LEARNING AND SIGMOIDAL OUTPUTS

Once we know that any decision rule can be approximated by a multilayer network, the next crucial question is "How can we get a network to *learn* a good decision rule?"

As before, we would like to use a set of training examples to come up with a choice of weights such that the network performs well on the training examples.

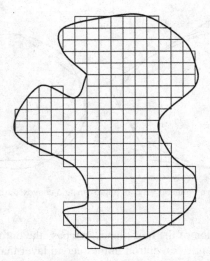

Figure 10.4   Approximating a nonconvex set with hypercubes.

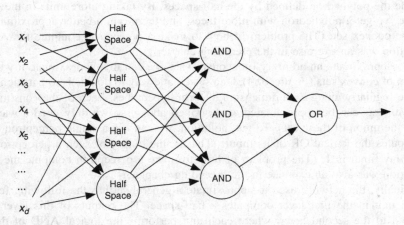

Figure 10.5   Network structure to approximate general sets.

Owing to the complexity of the network and the nonlinear relationships between the weights and the network output, a direct method of finding weights to make the network output fit the training data is out of the question. We need an incremental learning rule analogous to the perceptron convergence procedure to make gradual adjustments to the weights in a reasonable way.

We are faced with two difficult problems. In the case of a single unit, we exploited the simple relationship between each weight and the effect on the output to determine how we should adjust the weight. We no longer have such a simple relationship and will need to rely on some systematic analysis to help us determine

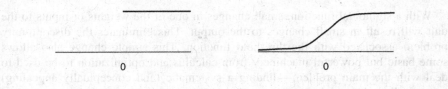

Figure 10.6 Threshold functions versus sigmoid functions.

reasonable adjustments to the weights. The second problem that further complicates the situation is caused by the nature of the output function of each unit. The sign($\cdot$) function is a type of "threshold" function that makes gradual tuning difficult.

Consider the effect of the discontinuity in the threshold function shown on the left in Figure 10.6. As we start making small changes to a particular weight, there will be no effect on the output if the total input to the unit is not close to zero (if we are far away from the discontinuity). However, once the total input to the unit is near zero, the output can change drastically with a tiny change in the weight. This happens when the change in the weight causes the input to the unit to cross the discontinuity of the threshold function.

Although this was not an issue in the case of a single perceptron, now outputs of some units are fed as inputs to others. Hence, a drastic change at the output of just one unit can create further drastic changes in many other downstream units, making an analysis of the effect at the final output difficult.

We can solve the problem of discontinuous threshold functions by making a modification in the output function of the units. Instead of using the threshold function sign($\cdot$), we use a "smoothed" version such as that shown on the right in Figure 10.6. It turns out that having the smooth function vary between 0 and 1 instead of -1 and 1 simplifies things. So, for convenience, we make this change as well, although this is not the main point. The key change is that the discontinuous threshold function is now replaced by a smoothly varying function. A function with this general "S" shape that is differentiable and increasing and tends to finite limits at $-\infty$ and ∞ is called a *sigmoid function*.

Thus, the output of a unit is now given by

$$a = \sigma(x_1 w_1 + \cdots + x_d w_d). \tag{10.1}$$

where the x_i are the inputs to the unit, the w_i are the weights, and $\sigma(\cdot)$ is a sigmoid function. One commonly used sigmoid function is

$$\sigma(y) = \frac{1}{1 + e^{-y}}. \tag{10.2}$$

Here e is the base of the natural logarithm, $e = 2.71828 \ldots$. The function $\sigma(y)$ ranges between 0 and 1 and has a derivative (or slope) given by

$$\sigma'(y) = \frac{d\sigma}{dy} = \sigma(y)(1 - \sigma(y)). \tag{10.3}$$

With a sigmoidal function, small changes in one of the weights or inputs to the unit will result in small changes to the output. This eliminates the discontinuity problem associated with the threshold function. This simple change also allows some basic but powerful machinery from calculus and optimization to be used to deal with the main problem—finding a systematic (and conceptually appealing) way to adjust the weights.

One final note on the implications of using a sigmoidal function instead of a threshold function concerns the final output of the network. With a sigmoidal function, the final output can take on arbitrary real values between 0 and 1 (instead of just two values, -1 and 1). This does not cause a problem either in training or in classification. During training, the target output can be taken to be 0 or 1 even though the actual output will be something in between. We can still measure error by considering the differences between t and a as discussed further below. Once a network has been trained and we wish to use the network for classification, we can simply decide class 1 if the output is greater than $1/2$ and decide class 0 otherwise. This is like passing the output of the network through a final threshold unit to get a binary classification, even though the rest of the network has sigmoidal units.

10.3 TRAINING ERROR AND WEIGHT SPACE

Because the network is now complicated, we cannot easily specify a set of weights such that the network output agrees with the target output for all of the training examples. Instead of directly specifying the weights, we use a sequential training method to try to come up with a set of weights that performs well on the training examples.

For each training example, the error of the network on the example is the difference $t - a$ between the target output and the actual network output. We wish to find a set of weights to make the error small over all the training examples. However, we do not want to simply sum the errors for each training example and minimize this sum, because then positive errors can cancel out negative errors. That is, we may end up with a set of weights such that the error is very large and positive for some training examples and is large and negative for others. In this case, the sum of the errors may be zero (or close to zero) even though this is certainly not a desirable situation.

To remedy this problem, we can instead sum the absolute values of the errors, $|t - a|$, or more commonly, sum the squares of the errors $(t - a)^2$ over all the training examples. The squared error is preferable to the absolute error because it is differentiable while the absolute value is not. This allows us to use the machinery of calculus as mentioned before and as we will shortly see.

One way to visualize this is to consider the error over the training examples as a function of the weights in the network. Each of the W weights in the network can be varied independently of the others, so the set of all possible weights can be thought of as a W-dimensional space. For each point in this weight space (i.e., for each choice of weights), we can compute the error of the corresponding network

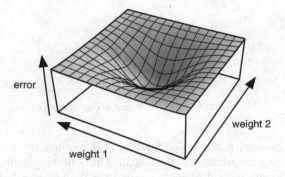

Figure 10.7   Error surface for 2D problem.

over the training examples. This results in an "error surface," and as we change the weights, we move around on this surface. A two-dimensional example is shown in Figure 10.7. Of course, most networks of interest have many more than two weights, so the "surface" is actually in a much higher dimensional space. Although this is difficult to depict, the higher dimensional case is conceptually similar to the two-dimensional case. Our goal is to find a choice of weights that minimizes the error. Equivalently, we wish to find weights for which the error surface is as low as possible, which corresponds to finding the deepest valley in the error surface.

10.4   ERROR MINIMIZATION BY GRADIENT DESCENT

If we had access to the complete error surface in a suitable form, we could simply choose those weights that give the minimum error. However, in practice the only knowledge we have about the error surface is its value at samples of those choice of weights for which we have actually taken the trouble to compute the error. Remember that to get this error, we need to check how the network performs over all the training examples.

With a large number of weights (which is the typical case), the weight space is so vast that it is infeasible to try to get a sense of the entire error surface by sampling it over a dense collection of weights. Instead, we have to settle for having information at a relatively small number of points. Moreover, the error surface is generally bumpy and complicated (with many hills and valleys), so a sparse sampling of weights often fails to give an idea of where the optimum weights lie.

This is an example of an extremely difficult optimization problem. For such problems, one often settles for a "local" minimum as opposed to the "global" minimum. That is, rather than trying to find the set of weights that gives the minimum error over all possible weights, one might try to simply find weights for which no improvement in the error can be obtained by very small changes in the weights. Using the error surface analogy, this is like trying to find the bottom of a valley, even if this valley may not be the deepest valley.

Figure 10.8   Gradient descent in one dimension.

A standard class of methods for finding local minima are known as descent algorithms. These are equivalent to the so-called hill-climbing methods, which seek to find local maxima instead of local minima (or the top of a hill instead of the bottom of a valley). If we preferred hill climbing to descent, we could simply try to maximize the negative of the error rather than minimize the error, thus converting a minimization problem into a maximization problem.

A popular descent algorithm is known as gradient descent. This algorithm can be applied when we know (or can get estimates of) the derivatives of the error surface with respect to the weights. The gradient is a vector in which the i-th component is the partial derivative with respect to the i-th weight. The partial derivative with respect to a weight gives the slope of the error surface in the direction of that particular weight. It turns out that the gradient gives the direction of maximum slope. This is the direction in which we should adjust the weights so that a small change in weights will give the largest change in error.

If we let E_m denote the error on the training example m, then on the basis of this training example, gradient descent suggests that we update each weight w_{ij} by the following amount (where η is a constant)

$$\Delta w_{ij} = -\eta \frac{\partial E_m}{\partial w_{ij}}. \tag{10.4}$$

Figure 10.8 depicts gradient descent in one dimension. The above equation makes sense conceptually, but we still need a way to compute the partial derivative to get an explicit expression for Δw_{ij} in terms of known quantities. A way to do this is provided by the backpropagation algorithm discussed in the next section.

10.5 BACKPROPAGATION

Backpropagation (or, backprop) gives a nice sequential implementation of gradient descent. For each training example, it provides a way to explicitly compute the weight adjustments of Equation (10.4). Because backprop is simply an implementation of gradient descent, it generally leads only to a local minimum. However, despite this seemingly serious limitation (of giving a local minimum rather than the global minimum) backpropagation has been found to give excellent results in many real applications. Various forms of backpropagation are by far the most widely used training algorithms for multilayer neural nets.

Backprop for a Single Unit

We first describe backpropagation in the case of a single unit. Consider a unit such as that shown in Figure 10.9. In this case, by evaluating Equation (10.4) in terms of known quantities, we get

$$\Delta w_i = \eta(t - a)\sigma' x_i. \tag{10.5}$$

Here σ' is the derivative of σ with the weighted sum of the inputs as argument. A derivation of this equation is given in the following section. Not surprisingly, the expression in Equation (10.5) is exactly the expression for Δw_i from the perceptron convergence procedure, but with the extra σ'. So, at least in the case of a single unit, backpropagation seems to be doing something reasonable.

The intuition discussed previously for the single unit case (for the perceptron convergence procedure) still applies, although now recall that t is either 0 or 1 (instead of -1 or 1), and that a is between 0 and 1. Suppose the target is 1 for a certain training example, but the output of the unit is less than 1. To get the output closer to 1, we would like to increase the total input to the unit. This can be achieved by increasing the weight w_i if the corresponding value x_i is positive and by decreasing the weight w_i if x_i is negative. Equation (10.5) does just that.

In the case of a sharp threshold unit, the unit output a (as well as the target output t) could be only two possible values. Thus, if the difference $t - a$ was nonzero, then it had to be either 2 or -2, and it was really the sign of the difference that was crucial in the weight change (since the magnitude could be absorbed in the constant). Now, the difference $t - a$ can be any value in the interval $(-1, 1)$, and so the magnitude (as well as the sign) of the term $t - a$ also plays a role. If $t = 1$ and a is close to 1, then the difference will be small and this will result in a smaller adjustment to the weight than if a were far from 1. It is reasonable that the adjustment should be greater if the output is further from the desired value.

Backprop for a Network

In the case of a general multilayer network, we will have a number of layers and a number of neurons in each layer. Each weight connects a unit from one layer to some unit in the next layer. To describe how a particular weight should be adjusted, we introduce some notation so that we can keep all the various neurons (and their inputs and outputs), weights, and layers straight.

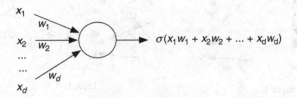

Figure 10.9   Single unit.

Consider a network as shown in Figure 10.10. There are L layers with layer 1 as the first layer (to which the original inputs $x_1, \ldots, x_d$ are connected) and layer L as the output layer (consisting of just a single neuron). Let l be an index that indicates which layer we are discussing, so that l takes on the values $1, 2, \ldots, L$.

We introduce the following notation for the input, output, and weights associated with various neurons:

$u_i(l) =$ total input to neuron i in layer l,

$a_i(l) =$ output of neuron i in layer l,

$w_{ij}(l) =$ weight from neuron j in layer $l-1$ to neuron i in layer l.

We can now give the equations for the backpropagation algorithm. We need to specify the quantity $\Delta w_{ij}(l)$ which tells how much we should adjust the weight $w_{ij}(l)$. The algorithm is specified by the equations

$$\Delta w_{ij}(l) = -\eta \; \delta_i(l) \; a_j(l-1), \tag{10.6}$$

where

$$\delta_i(l) = \sigma'(u_i(l)) \sum_k \delta_k(l+1)w_{ki}(l+1) \text{ for } l = 1, 2, \ldots, L-1 \tag{10.7}$$

and

$$\delta_i(L) = \sigma'(u_i(L)) \; (t_i - a_i(L)). \tag{10.8}$$

We have assumed that there is only a single neuron in the output layer, so that we could drop the subscript i in Equation (10.8). As written, the equations apply to the general case of multiple neurons in the output layer.

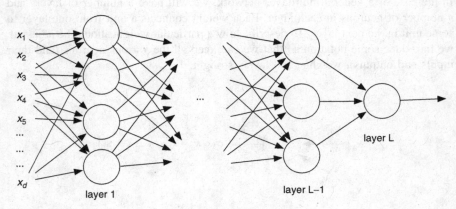

Figure 10.10 Network.

The most complicated term in the expression for $\Delta w_{ij}(l)$ is $\delta_i(l)$. This is a type of "error" term, and so the expression for $\Delta w_{ij}(l)$ is of the form

$$\Delta w_{ij}(l) = -\eta \cdot (\text{error term}) \cdot (\text{input term}). \tag{10.9}$$

which is of the same form as the expression for the single unit case.

Although the expression for $\delta_i(l)$ looks complicated, the first thing to notice is that, aside from the term $\delta_k(l+1)$, all the other terms are quantities that we can easily compute. The term $\delta_k(l+1)$ is just another error term for a downstream neuron—the k-th neuron in the next layer $l+1$. Moreover, we have a separate equation for $l = L$ that is simple and for which all the quantities are easily computed. This suggests that we start from the last layer (layer L which is the output layer) and work backwards.

This is precisely the reason for the term "backpropagation." In computing the output of the network, we start at the first layer and knowing the inputs $x_1, \ldots, x_d$, we propagate the computation forward until finally we reach the last layer and compute the output. But to compute the weight adjustments, we start at the output layer (layer L), and compute the error term first for this layer using Equation (10.8) and the input and output of the output neuron. We then propagate these error terms backwards through the network (layer by layer) using Equation (10.7), until we finish at the first layer. The backward propagation uses the inputs and outputs to various neurons that were computed on the forward pass. The weight adjustments are easily computed once we have the error terms (and the inputs computed on the forward pass).

Thus, the backpropagation algorithm can be summarized as follows:

1. Choose some initial values for all the weights and for the constant η.
2. Repeatedly cycle through the training examples. For each training example considered, do the following:
   (a) Compute the total input $u_i(l)$ for each neuron in each layer starting from the input layer and propagating forward to the output layer. Also, compute the final outputs of the network $a_i(L)$.
   (b) Starting at the output layer (layer L) and propagating backward to the second layer, compute the unit errors $\delta_i(l)$ using equations (10.8) and (10.7).
   (c) Add $\Delta w_{ij}(l)$ to the weight $w_{ij}(l)$, where the weight adjustments $\Delta w_{ij}(l)$ are given by equation (10.6).

10.6 DERIVATION OF BACKPROPAGATION EQUATIONS*

In this section, we give a derivation of the equations for the backpropagation algorithm (Equations (10.5) and (10.6)). As mentioned above, backprop is just an implementation of gradient descent, and so we only need to explicitly compute the partial derivative in Equation (10.4) in terms of known quantities. This will be done through a judicious use of the chain rule from calculus. The case of a single unit is particularly simple, and so we treat this case first.

10.6.1 Derivation for a Single Unit

Recall that when presented with the m-th training example, the gradient descent rule is to set the adjustment Δw_i to weight w_i to be

$$\Delta w_i = -\eta \frac{\partial E_m}{\partial w_i}. \tag{10.10}$$

The term $\frac{\partial E_m}{\partial w_i}$ measures how E_m varies as we vary w_i. As we change the weight w_i, all the variation in E_m is through the total input, u, to the unit. Thus, using the chain rule, we get

$$\frac{\partial E_m}{\partial w_i} = \frac{\partial E_m}{\partial u} \frac{\partial u}{\partial w_i}. \tag{10.11}$$

Recall that the total input to the unit is given by

$$u = x_1 w_1 + \cdots + x_d w_d. \tag{10.12}$$

Now, both terms on the right hand side of Equation (10.11) are easy to compute. From the expression for the input u, we see that

$$\frac{\partial u}{\partial w_i} = x_i. \tag{10.13}$$

To compute the first term, note that the error E_m is given by

$$E_m = \frac{1}{2}(t - a)^2 = \frac{1}{2}(t - \sigma(u)^2). \tag{10.14}$$

Therefore,

$$\frac{\partial E_m}{\partial u} = -(t - \sigma(u))\frac{\partial \sigma(u)}{\partial(u)} = -(t - a)\sigma'(u). \tag{10.15}$$

Using the expressions of Equations (10.15), (10.13), and (10.11) in Equation (10.10), we get

$$\Delta w_i = \eta(t - a)\sigma'(u)x_i, \tag{10.16}$$

which is the result stated in Equation (10.5).

10.6.2 Derivation for a Network

We can now derive the backpropagation algorithm in the case of a general multi-layer network. Recall that $u_i(l)$ and $a_i(l)$ are, respectively, the total input to and output of neuron i in layer l, and $w_{ij}(l)$ is the weight from neuron j in layer $l-1$ to neuron i in layer l.

Notice that the input $u_i(l)$ to neuron i in layer l comes from the output of neurons in the previous layer $l-1$. These outputs get multiplied by the corresponding weights and added up to form $u_i(l)$. Therefore,

$$u_i(l) = \sum_j w_{ij}(l)a_j(l-1). \tag{10.17}$$

The output of this neuron is then just $u_i(L)$ passed through the sigmoid function. Namely,

$$a_i(l) = \sigma(u_i(l)) = \sigma\left(\sum_j w_{ij}(l)a_j(l-1)\right). \tag{10.18}$$

As before, gradient descent suggests that on training example m we should adjust $w_{ij}(l)$ by

$$\Delta w_{ij}(l) = -\eta \frac{\partial E_m}{\partial w_{ij}(l)} \tag{10.19}$$

and the only question is how to compute this effectively.

Notice that the final output of the network, $a(L)$, depends on the weight $w_{ij}(l)$ only through the input to the i-th neuron in layer l, $u_i(l)$. Hence, using the chain rule, we obtain

$$\frac{\partial E_m}{\partial w_{ij}(l)} = \frac{\partial E_m}{\partial u_i(l)} \frac{\partial u_i(l)}{\partial w_{ij}(l)}. \tag{10.20}$$

This is analogous to Equation (10.11) for the single unit case.

The second term is easy to compute. Using the expression (10.17), we get

$$\frac{\partial u_i(l)}{\partial w_{ij}(l)} = a_j(l-1). \tag{10.21}$$

For the first term, we introduce the notation

$$\delta_i(l) = -\frac{\partial E_m}{\partial u_i(l)}. \tag{10.22}$$

To finish the derivation, we only need to show the recursive equation for computing $\delta_i(l)$ given by Equations (10.7) and (10.8).

First, consider $l = L$ (the output layer). Because this is the output layer, E_m can be written easily in terms of $u_i(L)$, just as in the single unit case. Specifically,

$$E_m = \frac{1}{2}(t - a_i(L))^2 = \frac{1}{2}(t - \sigma(u_i(L))^2. \tag{10.23}$$

Therefore,

$$\delta_i(L) = -\frac{\partial E_m}{\partial u_i(L)} = (t - a(L)) \, \sigma'(u_i(L)), \tag{10.24}$$

which is the result given in Equation (10.8).

Now, consider $l < L$. For this case, E_m cannot be written easily in terms of $u_i(l)$. However, note that E_m depends on $u_i(l)$ only through the output $a_i(l)$ of the i-th neuron in layer l. So, using the chain rule once again, we get

$$\delta_i(l) = -\frac{\partial E_m}{\partial a_i(l)} \frac{\partial a_i(l)}{\partial u_i(l)}. \tag{10.25}$$

Since $a_i(l) = \sigma(u_i(l))$, we have

$$\frac{\partial a_i(l)}{\partial u_i(l)} = \sigma'(u_i(l)). \tag{10.26}$$

To complete the derivation, we only need an expression for $\partial E_m/\partial a_i(l)$. To this end, we make a final use of the chain rule, noting that E_m depends on $a_i(l)$ only through inputs to neurons in the next layer (i.e., layer $l + 1$). The output $a_i(l)$ may affect the inputs to all the neurons in layer $l + 1$, so we need many terms, one for each $u_k(l + 1)$ over all neurons k. Hence,

$$-\frac{\partial E_m}{\partial a_i(l)} = -\sum_k \frac{\partial E_m}{\partial u_k(l + 1)} \frac{\partial u_k(l + 1)}{\partial a_i(l)}. \tag{10.27}$$

By the definition of the error term introduced in Equation (10.22), we have

$$-\frac{\partial E_m}{\partial u_k(l + 1)} = \delta_k(l + 1). \tag{10.28}$$

And, by the expression for the total input to neuron k in layer $l + 1$, we have

$$\frac{\partial u_k(l + 1)}{\partial a_i(l)} = \frac{\partial}{\partial a_i(l)} \left[\sum_j w_{kj}(l + 1)a_j(l) \right] = w_{ki}(l + 1). \tag{10.29}$$

Thus,

$$-\frac{\partial E_m}{\partial a_i(l)} = -\sum_k \delta_k(l + 1)w_{ki}(l + 1) \tag{10.30}$$

so that

$$\delta_i(l) = \sigma'(u_i(l)) \sum_k \delta_k(l+1) w_{ki}(l+1), \qquad (10.31)$$

which is the result of Equation (10.7).

10.7 SUMMARY

In this chapter, we continued our discussion of neural net learning, this time discussing feedforward networks with multiple layers of nodes. We modified the simple sharp threshold function used in a simple perceptron to a sigmoid function that always has a slope. We explained how the backprop algorithm can be used to train such a network by changing the weights of connections between neurons. Backprop is a sequential implementation of a gradient descent type algorithm. Although this results in weights that are locally optimal (instead of globally optimal), the algorithm has been found to work well in practice and is the most widely used training algorithm for multilayer networks.

10.8 APPENDIX: GRADIENT DESCENT AND REASONING TOWARD REFLECTIVE EQUILIBRIUM

Human reasoning can be viewed as a process of changing one's view by addition and subtraction, that is, adding new beliefs and removing some prior beliefs. People make such changes in order to increase the "coherence" of their overall views. Coherence can be increased by giving up beliefs that are in tension with other beliefs and by adding new beliefs that fit with or help to explain other beliefs. People try to make small changes that will improve overall coherence.

This can be thought of as a kind of gradient descent: making small changes that will reduce the incoherence of one's overall beliefs. Or gradient ascent: making small changes that increase the overall coherence of one's overall beliefs.

Philosopher John Rawls describes this as trying to reach what he calls a "reflective equilibrium," in which one's beliefs fit together in a coherent way. (This is in his book, *A Theory of Justice*.)

It is sometimes objected that this can be an overly conservative way to proceed and that it is sometimes better to try out more radical ideas to see whether larger changes do better at reducing incoherence (or increasing coherence). This is sometimes called trying to reach a "wide reflective equilibrium." The idea is perhaps that the more conservative method runs the risk of getting stuck in a local minimum (or maximum), something that might be avoided by being bolder.

Suppose you are trying to reach a "wide reflective equilibrium" in your beliefs about morality, by considering whether your general moral principles match up with your beliefs about particular cases. You may discover conflicts between your beliefs; for instance, suppose you thought that it is always wrong to steal, but this conflicts with your belief that it is morally permissible for a starving person to steal a loaf of bread. You revise your general moral principle to the claim that it is wrong to steal unless doing so will save someone's life. This new principle conflicts with further beliefs that you have, for example, that it is ok for a hungry child to steal food even if the child is not about to die. So you revise your general principle further to accommodate this case, and so on.

10.9 QUESTIONS

1. (a) What is a single perceptron?
   (b) What is a multilayer feedforward neural network?

2. Consider the XOR problem with two inputs x_1 and x_2. That is, the ouput is 1 if exactly one of x_1, x_2 is positive, and the output is 0 otherwise. Construct a simple three-layer network to solve this problem.

3. What is a convex set?

4. Explain the sense in which for any decision rule there is a three-layer network that approximates that rule. Sketch a proof of this.

5. Consider a classification problem in which each instance consists of d features $x_1, \ldots, x_d$, each of which can only take on the values 0 or 1. A feature vector belongs to class 0 if $x_1 + x_2 + \cdots + x_d$ is even (i.e., the number of 1's is even) and it belongs to class 1 otherwise. Can this problem be solved by a single perceptron? Why or why not? A three-layer network?

6. True or false: The backpropagation learning method requires that the units in the network have sharp thresholds.

7. True or false: The backpropagation learning method for feedforward neural networks will always find a set of weights that minimizes the error on the training data.

8. What is "gradient descent" and what is a potential problem with it?

9. Why is the weighted input to a unit in a feedforward network passed through a sigmoid function rather than a simple threshold function?

10. Explain why the sort of learning rule discussed in this chapter is appropriately called "backpropagation."

11. Is the process of reasoning toward reflective equilibrium analogous to the algorithm of gradient descent that is used to train a neural network?

10.10   REFERENCES

In the 1960s, along with the interest in perceptrons, there was also interest in multilayer networks. However, an adequate training rule was lacking. The first backpropagation-like algorithm appears to be from Werbos (1974), though it only became widely known after the work of Rumelhart *et al.* (1986a,b), and through the book *Parallel Distributed Processing* edited by Rumelhart and McClelland (1986). This and other work in the mid 1980s sparked great interest in neural networks that has continued to date. Most recent books that discuss perceptrons also discuss multilayer perceptrons and backpropagaton. In particular, Mitchell (1997), Russell and Norvig (2010), Duda *et al.* (2001), and Haykin (1994) all provide additional reading for the topics of this chapter as well.

Rawls (1971) discusses reasoning to a "reflective equilibrium." Harman (1986) says more about such reasoning.

Duda RO, Hart PE, Stork DG. Pattern classification. 2nd ed. New York: Wiley; 2001.

Harman G. Change in view: principles of reasoning. Cambridge (MA): MIT Press; 1986.

Haykin S. Neural networks: a comprehensive foundation. New York: Macmillan Publishing Company; 1994.

Mitchell TM. Machine learning. Instance-based learning. Boston (MA): McGraw-Hill; 1997. pp. 226–229, Chapter 8.

Rawls J. A theory of justice. Cambridge (MA): Harvard University Press; 1971.

Rumelhart DE, McClelland JL, editors. Volume 1, Parallel distributed processing: explorations in the microstructure of cognition. Cambridge (MA): MIT Press; 1986.

Rumelhart DE, Hinton GE, McClelland JL. Learning representations by back-propagating errors. Nature 1986a;323:533–536.

Rumelhart DE, Hinton GE, McClelland JL. Learning internal representations by error propagation. In: Rumelhart DE, McClelland JL, editors. Volume 1, Parallel distributed processing. Cambridge (MA): MIT Press; 1986b, Chapter 8.

Russell S, Norvig P. Artificial intelligence: a modern approach. Artificial neural networks. Upper Saddle River (NJ): Prentice-Hall; 2010. pp. 563–597, Chapter 18.

Werbos PJ. Beyond regression: new tools for prediction and analysis in the behavioral sciences [PhD thesis]. Harvard University, Cambridge (MA); 1974.

CHAPTER 11

PAC Learning

So far we have discussed two rather different approaches to learning pattern recognition. The first approach is illustrated by nearest neighbor rules. By taking more and more neighbors, but in such a manner that the number of neighbors grows more slowly than the number of training examples, we are guaranteed that as the amount of data grows we will do in the limit as well as the optimal Bayes decision rule. A similar result holds for kernel rules.

The second approach we discussed uses neural networks. By taking a multilayer perceptron with enough layers and enough nodes per layer, we can approximate any decision rule arbitrarily closely. We also described the backpropagation training algorithm, which, although guaranteed to converge only to a local minimum, works very well in practice. Of course, if we are willing and able to expend a sufficient amount of computing power, we can ensure that we arrive at the best possible weights for classifying the training examples, rather than just a local optimum. However, even in this case, we do not yet have any guarantee that the network will generalize well on *new* examples. In any case, neural nets provide another approach to learning problems.

One thing that we have been lacking so far is some measure of the inherent difficulty of a learning problem. That is, we would like some way to make statements about which problems have solutions that are learnable by some algorithm or other—not just statements regarding the performance of a *particular* learning method. In addition to knowing whether or not a problem has "learnable" solutions, we would like some measure of the inherent difficulty of the learning problem. Some early work in this direction was done in the probability and statistics community. There has been continued work along these lines by researchers in a variety of fields, including statistics, electrical engineering, and computer science. This chapter begins a discussion of some of the results that have been obtained in this area.

An Elementary Introduction to Statistical Learning Theory, First Edition.
Sanjeev Kulkarni and Gilbert Harman.
© 2011 John Wiley & Sons, Inc. Published 2011 by John Wiley & Sons, Inc.

11.1   CLASS OF DECISION RULES

Recall that a decision rule has to choose either "decide 0" or "decide 1" whenever presented with an observed feature vector $\bar{x}$. Therefore, a decision rule carves the feature space into two sets—those feature vectors for which the rule decides 0 and those for which it decides 1. We can represent a decision rule as a map c from the feature space, say $\mathbf{R}^d$, to $\{0, 1\}$, with $c(\bar{x})$ representing the decision when we observe feature vector $\bar{x}$. Equivalently, we can think of a decision rule as a subset of the feature space, associated with those feature vectors for which the rule decides 1.

In formulating a definition of learnability, we suppose that the learner's task is to select some member of a certain class C of decision rules. There are several reasons for specifically mentioning the class C of decision rules. In certain problems, there may be design or structural constraints on the learner that allow the use of only certain restricted forms of rules. Computational issues may impose constraints on the feasibility of working with particular classifiers. Yet another reason to work with a specific class of decision rules is to exploit some prior knowledge associated with the problem domain.

In addition to the practical reasons for working with a fixed collection of decision rules, this perspective also leads to a fruitful conceptual framework. The results obtained provide a characterization of the difficulty of a learning problem in terms of a measure of the "richness" of the class of decision rules.

If there is only one possible decision rule the learner can select (so that the class C has only one member), then "learning" is a nonissue. In this case, the learner has no choice but to select this one decision rule, regardless of any data observed.

The issue of learning arises when the learner must select a decision rule from a collection C of possible rules under consideration. The learner uses labeled examples and possibly other prior information to make this selection. In this case, there is uncertainty at the outset regarding which rule from the class is best. After making observations, the uncertainty regarding the best rule from the class is presumably reduced. Learning corresponds to improved performance in selecting a relatively good rule on the basis of observations.

Assuming that the learner selects a decision rule from some fixed class of rules may seem artificial and restrictive. However, it is actually quite general and most standard learning algorithms can be framed in this way. Of course, one could just take C to be the set of *all* decision rules. Unfortunately, as we will see in Chapter 12, this degenerate way of framing a problem in terms of a class of all decision rules is not very useful, since the class is too rich to be learnable in any reasonable way. But many interesting learning problems result in nondegenerate classes of decision rules for which useful results can be obtained.

For example, suppose we are interested in learnability using a multilayer perceptron. Consider a network with a fixed architecture—a fixed number of nodes in each layer and a fixed number of connections between the nodes. For a given set of weights, the network computes the corresponding decision rule. Changing the weights causes the network to compute a different rule. We can consider the set of all possible rules the network can represent as we change the weights over all possible values. This set of rules determines the class C. In learning (e.g., using backpropagation), the choice of weights is determined from a set of training examples. This results in selecting one of the rules from the set C. In this case, the results from Chapter 12 can be applied directly.

In the case of nearest neighbor methods, the class of decision rules consists of the possible labelings (0 or 1) of all possible Voronoi regions deriving from any finite number of data points. This situation allows for treatment along lines we discuss in Chapter 13.

11.2   BEST RULE FROM A CLASS

We assume the setup described in Chapter 4 in the formulation of the pattern recognition problem. That is, we assume that there are two classes 0 and 1 with prior probabilities $P(0)$ and $P(1)$, respectively. The feature vectors are related to the class of the object through conditional distributions $P(\overline{x}|0)$ and $P(\overline{x}|1)$ or conditional densities $p(\overline{x}|0)$ and $p(\overline{x}|1)$.

Suppose a class C of decision rules is fixed. These are the only rules that we can use for classification. We see labeled examples $(\overline{x}_1, y_1), \ldots, (\overline{x}_n, y_n)$ and wish to select a good decision rule from the class C.

An immediate question that arises is what is meant by a "good" decision rule? We answered this question in one context in Chapter 5. There we argued that if the distributions are known, then the optimal decision rule is the corresponding Bayes rule, which has an error rate denoted by R^*.

However, as before, we assume that the distributions are not known, so that we do not have the information necessary to compute Bayes rule. Instead, we need to find a good rule only on the basis of the data $(\overline{x}_1, y_1), \ldots, (\overline{x}_n, y_n)$ and the class C.

We might try to achieve a performance close to that of the Bayes rule, as we did with the nearest neighbor rule. Unfortunately, given that we are restricted to using rules from C, we may not always be able to find a classifier with performance close to a Bayes rule. It just might happen that there are no rules in C with error rate close to R^*.

So, what is the best error rate we can hope for? Each rule $c \in C$ has an associated error rate

$$R(c) = P(0)P(\Omega_1(c)|0) + P(1)P(\Omega_0(c)|1),$$

where $\Omega_0(c)$ is the set of feature vectors for which c decides 0 and $\Omega_1(c)$ is the set of feature vectors for which c decides 1. In other words, the error rate for c is the sum of the probability that a randomly encountered object belongs to class 0 but c decides 1, and the probability that a random object belongs to class 1 but c decides 0.

A natural quantity to consider is the minimum error rate over all rules from C, which we denote by R_C^*. That is,

$$R_C^* = \min_{c \in C} R(c).$$

Strictly speaking, one should replace the min with inf, the lower limit of the error rate for hypotheses in c, which allows dealing with cases where no rule from the class actually achieves the minimum, but rules in the class do get arbitrarily close to it. R_C^* is the best performance we can hope for if we are restricted to using rules from C.

Note that we always have $R_C^* \geq R^*$, since the Bayes rate is the best possible over all decision rules. We certainly cannot do better by restricting the class of rules. However, if C happens to contain a rule close to Bayes rule, then R_C^* may be close to the Bayes rate R^*, or we may even have $R_C^* = R^*$.

It is also important to note that we cannot actually compute the error rates for the rules in C, since we do not know the necessary probability distributions. If we were able to do so, then learning would not be an issue. We would simply compute the error rate of each rule in C and select the one with best performance. In fact, if this were the case, we could even just go ahead and compute the optimal Bayes decision rule. Instead, we have to settle for trying to select a good rule from C on the basis of the data. This is the learning problem.

11.3 PROBABLY APPROXIMATELY CORRECT CRITERION

We argued in the previous section that by requiring our decision rules to belong to the class C, we can only strive for rules with error rate R_C^* instead of R^*. In this section, we discuss why even this goal should be relaxed somewhat.

We are interested in how much data we need for learning, or equivalently, how well we can do with a given finite amount of data. On the basis of some training examples $(\overline{x}_1, y_1), \ldots, (\overline{x}_n, y_n)$, our task is to produce a hypothesis $h \in C$. With a finite amount of data, it is unreasonable to expect to be able *always* to find the best rule from C, and so we should not expect to be able to produce a hypothesis with an error rate that exactly equals R_C^*. Instead, we will settle for a rule that is only approximately optimal (or *approximately correct*). That is, we would like to select a hypothesis h from C such that the error rate $R(h)$ of the hypothesis satisfies $R(h) \leq R_C^* + \epsilon$, where ϵ is an accuracy parameter.

Moreover, since the training examples are random, there is a chance that we will be unlucky and see poor or atypical examples. Therefore we should not even expect *always* (i.e., for *any* set of observed examples) to produce an h that is approximately correct. Rather, we will require only that we produce a good hypothesis with high probability. In other words, we will require that

$$P\{R(h) \leq R_C^* + \epsilon\} \geq 1 - \delta$$

for some confidence parameter δ. This is the *Probably Approximately Correct* (PAC) criterion.

Although the name "Probably Approximately Correct" may sound funny, it is actually quite descriptive. By "correct" we mean producing the best possible hypothesis, that is, a hypothesis with the best possible error rate R_C^*. We argued

that with a finite number of training examples, we cannot always expect to produce
the correct hypothesis exactly, so we settle for one that is approximately correct.
Moreover, with random examples, we argued that we should not always expect to
be approximately correct with certainty. Rather we should expect only to probably
produce a hypothesis that is approximately correct.

Note that probabilities are actually computed twice in the PAC condition. The
inner, and perhaps less obvious, probability is inherent in the definition of $R(h)$.
This is an error rate, or probability of error, for the decision rule h. The rule h
classifies some feature vectors as 0 and others as 1. $R(h)$ measures the probability
that the decision rule h will make a mistake in classifying a feature vector. This
probability is over a *new* randomly encountered feature vector. If this probability
is small (say less than $R_C^* + \epsilon$), it means that we have come up with a good
hypothesis—in fact within ϵ of the best rule in C.

The outer (and explicit) probability in the PAC criterion, says that the probability
that we produce a good hypothesis is large. This probability is over the set of
training examples we see. We hope that the training examples are representative so
that a decision rule selected using the training examples will be a good one. If the
outer probability is large, it means that it is very likely (probability greater than
$1 - \delta$) that we see representative training examples.

The PAC criterion can be written in an equivalent way as

$$P\{R(h) > R_C^* + \epsilon\} < \delta.$$

Written this way, the event within the braces is the event that the hypothesis
produced is bad (has a high error rate). The PAC condition says that the probability
of producing a bad hypothesis (error rate larger than $R_C^* + \epsilon$) should be small (less
than δ). In the cases where a bad hypothesis is produced, all bets are off. There
are no guarantees on the error rate. A decision rule in these cases could even have
an error rate of 1, which is the worst possible since it means that the decision for
any feature vector is *always* wrong; or it could have an error rate of 1/2, which
is no better than ignoring the data and guessing randomly. The PAC criterion just
guarantees that these bad situations are not very likely.

11.4 PAC LEARNING

Now that we have the PAC criterion, we are almost ready to give a definition for
learnability, but there is one more issue to consider. This issue concerns the number
of training examples that are needed.

In the PAC criterion, we introduced the two parameters ϵ and δ. ϵ is an accuracy
parameter and δ is a confidence parameter. These two parameters control how
stringent we wish to make the PAC criterion.

Of course, we should expect that the number of examples we need to learn will
depend on the choice of ϵ and δ. However, we should also allow the number of
needed training examples for (ϵ, δ) learning to be independent of the underlying

distributions governing the problem. Recall that these distributions are the prior probabilities $P(0)$ and $P(1)$ and the conditional densities $p(\overline{x}|0)$ and $p(\overline{x}|1)$, or equivalently, $p(\overline{x})$ and $P(0|\overline{x})$, $P(1|\overline{x})$.

Remember that we have been assuming that we know little or nothing about these distributions, as this is precisely where learning comes in. Also, our goal at this point is to try to get some insight on how much data are required for learning. Hence, it is reasonable to require that the number of training examples required by a learning algorithm not depend on quantities unknown to the learner.

To clarify this point further, what we require is that given ϵ and δ, there is some finite sample size, which we denote by $m(\epsilon, \delta)$, such that $m(\epsilon, \delta)$ samples are enough for PAC learning, no matter what the underlying distributions happen to be. This does *not* mean that for certain distributions, we *must* see $m(\epsilon, \delta)$ training examples before a good hypothesis will be produced. On the contrary, if the distributions are degenerate and if we use a particular learning algorithm, then in some cases it may happen that a good hypothesis is achieved with just one example.

What is required is that no matter what the distributions are, we are guaranteed that some finite number of examples $m(\epsilon, \delta)$ will work for PAC learning to accuracy ϵ and confidence δ. This is distribution-free learning in the sense that nothing is assumed about the distributions. Some number of examples $m(\epsilon, \delta)$ is to be sufficient for all distributions. This sort of learning paradigm is also sometimes referred to as uniformly consistent learning, since the same number of examples is required to work (uniformly) for all distributions.

We also require that there is a function $m(\epsilon, \delta)$ that determines a finite sample size for every choice of ϵ and δ. That is, we do not want to specify ahead of time the accuracy and/or confidence with which we wish to learn. Instead, we want to guarantee that we can learn to whatever accuracy and confidence we choose. Of course, the number of training examples we need will normally depend on our choice of ϵ and δ.

We now define this notion of learnability, specifically, PAC learnability. The key ingredient is the class of decision rules C, and so the definition is for the learnability of the class C. Formally, we have the following definition.

PAC Learnability We say that a class of decision rules C is PAC learnable if there is a mapping that produces a hypothesis $h \in C$ based on training examples such that for every $\epsilon, \delta > 0$ there is a finite sample size $m(\epsilon, \delta)$ such that for any distributions we have

$$P\{R(h) > R_C^* + \epsilon\} < \delta,$$

after seeing $m(\epsilon, \delta)$ training examples.

Now the questions we are interested in are, "When is a class C PAC learnable?" and "If C is PAC learnable, can we get bounds on the number of examples $m(\epsilon, \delta)$ needed?" These questions are addressed in the next chapter.

11.5 SUMMARY

In this chapter, we discussed the use of a fixed class of decision rules. If we can use only the rules from some class C, we need to back off from the goal of trying to achieve the Bayes error rate R^*. Because we have only a finite amount of random data, we successively argued that we need to weaken the goal to R_C^*, to $R(h) < R_C^* + \epsilon$, and finally to $P\{R(h) < R_C^* + \epsilon\} > 1 - \delta$. This last condition is called the PAC criterion, since the hypothesis must be probably approximately correct. We then discussed the requirement that the sample size needed for ϵ, δ learning should naturally depend on ϵ and δ, but should be independent of the underlying distributions characterizing the problem. This led to the definition of PAC learnability of a class of rules C. In the following chapter, we discuss a characterization of those classes that are PAC learnable, and discuss the number of examples needed for ϵ, δ learning.

11.6 APPENDIX: IDENTIFYING INDISCERNIBLES

We have been concerned mainly with cases in which the connection between the observable features of an object and its correct classification are at least sometimes merely probabilistic. In such cases, the Bayes error rate cannot be zero.

But suppose we know that the correct classification of an object is completely determined by its observable features. Then, the Bayes error rate is zero, because there is a function from the features of the object to its correct classification, and so the error rate of that function is zero.

For that special case, we might *identify* objects with their features. The feature space then represents the set of all possible objects of interest and we can model a *concept* as that subset of objects (feature vectors) that are instances of the concept. These are the "positive instances" of the concept. Those objects (feature vectors) that are not in the subset are the "negative instances" of the concept. In the more usual case in which feature vectors do not always determine the correct classification, the set of positive instances of a concept cannot be identified with a set of feature vectors.

There are deep philosophical issues about whether different things could have exactly the same features. One such issue is whether there could be a universe with two things that are exactly alike—perhaps identical twins with exactly the same experiences and thoughts. It might be argued that they could not have exactly the same properties. For example, X might be on the right side of Y and Y on the left side of X; so that one has the feature of being on the right of someone otherwise the same, a feature the other does not have. But what if they are standing next to each other but facing in opposite directions, so that each is on the other's right side? (Think about it.) It might be argued that one would have to have been born first and the other second, so that would be a difference. But what if the twins are the simultaneous results of duplicating an original. It might be argued that one would be to the north of the other and the other to the south of the other. But suppose this occurs in space where there is no north and south. Or perhaps we can imagine a universe with two exactly similar spheres of gold and nothing else in it.

A related issue concerns the relation between a clay statue and the clay from which it has been made. It may seem that these must be two different things with exactly the same properties. But the statue has the property of having been made from the clay and the clay does not have the property of having been made from the statue. Are you the same thing as your body? Do you and your body have different properties? When you die, the body may still be there (if your death is not the result of a large explosion). So maybe your body has the property of *being able to exist even when you no longer exist*, a property you do not have.

11.7 QUESTIONS

1. What is the class of rules that can be represented by a perceptron?

2. Let C be a class of decision rules and let h be the hypothesis produced by a learning algorithm. Write the condition on the error rate $R(h)$ for PAC learnability in terms of R_C^*, ϵ, and δ.

3. One might argue that every learning algorithm works with a fixed set of decision rules, that is to say, the set of all rules that the particular algorithm might possibly produce over all possible observations. In light of such an argument, is there really anything new to the perspective of working with a fixed collection of decision rules?

4. If one learning algorithm works with a class of rules that includes all the rules of a different algorithm and some others, does this mean that the former algorithm should give a strictly better performance in a learning task? Why or why not? Discuss any advantages/disadvantages to be had by restricting the set of decision rules.

5. (a) Describe as precisely as you can what it means for C to be PAC learnable, explaining the roles of ϵ and δ and the requirements on the sample size.
   (b) Why do we settle for R_C^* instead of R^* and why do we introduce ϵ and δ?

11.8 REFERENCES

Working with a fixed class of decision rules in pattern recognition has a long history, including the focus on linear rules as early as the 1960s. Some of the key results on uniformly consistent learning for a general class appear to have started with the work of Vapnik and Chervonenkis (1971). The paper by Valiant (1984) spawned a great deal of interest on these topics in computer science, adding to the ongoing work in the probability, statistics, and pattern recognition communities. The term PAC learning originated and is mostly used in the computer science literature, while other terms for this approach are uniformly consistent estimation and VC theory. There are a number of books and review articles that discuss PAC learning. See, for example, Kearns and Vazirani (1994), Devroye *et al.* (1996), Vapnik (1996), Vidyasagar (1997), Kulkarni *et al.* (1998), and Anthony and Bartlett (1999).

Identity of indiscernables is discussed with references in Della Rocca (2005).

Anthony M, Bartlett PL. Neural network learning: theoretical foundations. Cambridge: Cambridge University Press; 1999.

Della Rocca M. Two spheres, twenty spheres, and the identity of indiscernibles. Pac Philos Q 2005;86:480–492.

Devroye L, Györfi L, Lugosi G. A probabilistic theory of pattern recognition. New York: Springer Verlag; 1996.

Kearns MJ, Vazirani UV. An introduction to computational learning theory. Cambridge (MA): MIT Press; 1994.

Kulkarni SR, Lugosi G, Venkatesh S. Learning pattern classification—A survey. IEEE Trans Inf Theory 1998;44(6):2178–2206.

Mitchell TM. Machine learning. New York: McGraw-Hill; 1997.

Russell S, Norvig P. Artificial intelligence: a modern approach. Learning from examples. Upper Saddle River (NJ): Prentice-Hall; 2010. pp. 645–767, Chapter 18.

Valiant LG. A theory of the learnable. Commun ACM 1984;27(11):1134–1142.

Vapnik VN. The nature of statistical learning theory. New York: Springer-Verlag; 1996.

Vapnik VN, Chervonenkis A. On the uniform convergence of relative frequencies of events to their probabilities. Theory Probab Appl 1971;16(2):264–280.

Vidyasagar M. A theory of learning and generalization, London: Springer-Verlag; 1997.

CHAPTER 12

VC Dimension

In the previous chapter we discussed a precise formulation of PAC learnability. Given a class of rules arising in some application, one could try to analyze the PAC learnability of the class by using the definition directly. However, instead of starting from scratch with the definition for each new situation, it is possible to provide a general characterization of those classes that are learnable. In addition, it is possible to say something about how much data are needed for learning in terms of the accuracy (ϵ) and confidence (δ) parameters. In this chapter we discuss a key result of this type for the PAC learning problem.

12.1 APPROXIMATION AND ESTIMATION ERRORS

Recall that in the PAC formulation, the learner is restricted to using decision rules from some class of rules C. There is an inherent trade-off involved in the richness of this class C.

On the one hand, we would like C to be extremely rich, that is, to contain lots and lots of rules. That way, the fact that we are required to use rules from C will not be a real limiting factor. In fact, if C is rich enough (e.g., if it contains or can approximate all possible decision rules), then we can even be sure that there are rules in the class that are as close as we wish to the optimal Bayes decision rule.

On the other hand, remember that we do not know which rule in C is a good one since we do not know the distributions characterizing the problem. We need to choose a good rule on the basis of the training examples. Clearly, trying to find out which rule from C is best on the basis of some training examples is more difficult if C is very rich. For example, in the degenerate case where C contains

An Elementary Introduction to Statistical Learning Theory, First Edition.
Sanjeev Kulkarni and Gilbert Harman.
© 2011 John Wiley & Sons, Inc. Published 2011 by John Wiley & Sons, Inc.

just one rule, picking the best rule from C is trivial, since there is only one rule to pick! Of course, as long as there is more than one rule, the training data must be used to decide which rule from C is likely to be the best in C. Hence, from this perspective we would like C to contain few rules since this would make the selection problem easier.

On the basis of the data, let us select a hypothesis $h \in C$. From the above discussion, we see that the performance of h is limited by two factors: (i) there simply may not be very good rules in C and (ii) we may be unable to identify a relatively good rule out of all the other rules in C using only the training data. Departure from the optimal Bayes rule due to the first cause is often called the *approximation error*, since it is caused by the inability to approximate Bayes rule using only rules from the class C. Our inability to select the best rule from the C is called the *estimation error* since it is caused by estimating the true error rates of rules in C using only the training data.

Now, we are considering the case where the class C is fixed, and we are interested in choosing good rules from C on the basis of the data. Thus, our focus in this and in the previous chapter is on the estimation error. Given a certain number of training examples, we would like to know how well we can do in terms of selecting a good rule from C. We know that the richness of C is important, but how do we characterize this richness? For example, is it just the number of rules in C that matters, or is it something else? We answer these questions in the next few sections.

12.2 SHATTERING

If C contains only a finite number of rules, then results on the estimation error as a function of the number of training examples can be obtained by some simple probabilistic arguments. However, if the number of rules in C is large, then these results are not so useful. Moreover, if C contains infinitely many rules, then the results are downright useless. Simply counting the number of rules in C is not the right measure of complexity (or richness) of C. A good measure of richness should take into account the "expressive power" of the rules in C. The notion of "shattering" is one way to capture the expressive power. Before giving a precise definition, we first try to motivate this notion with an example.

Suppose someone tells us that before the start of each week they can predict whether the stock market (say the S&P 500 index) will end the week higher or lower. In order to make this prediction, they measure various features such as the price behavior of the index over the previous several weeks, the recent behavior of interest rates and other financial and economic indicators, perhaps some company specific features such as earnings, and possibly others (such as whether the AFC or NFC won the Super Bowl that January, which has been playfully suggested to predict stock market performance for the year).

If we know that they use a fixed decision rule and find that they make correct decisions for 10 weeks straight without yet making an error, we may be rather impressed and place some confidence in their decision rule. If the "winning streak" of correct redictions continues for 52 weeks, we would be extremely impressed (and also possibly wealthy if we had invested according to the predictions). The chance that a random decision rule could match the outcomes for 52 straight weeks is extremely small (one in 2^{52}), so we would be quite confident that they are really on to something with the rule they have come up with. Even with just 10 straight correct predictions, the chance a random rule would achieve this performance is one in 1024.

On the other hand, suppose they tell us that instead of using a fixed decision rule, they have a collection of possible decision rules they may use. After the 10 weeks they search through their collection of decision rules and find one that agrees with the outcome for all the 10 weeks. Should we be impressed? Well, that should depend greatly on how many rules are in their class. If they have 1024 rules in their class and each of the possible 10-week outcomes is predicted by one of the rules, we should not be impressed at all. Of course one of the rules will agree, no matter what the results!

To be more precise, what really matters is not how many rules are in the class, but rather how many of the possible 10-week outcomes are represented by rules from the class. For example, even if there are thousands of rules in the class, but each one gives the same predictions over the 10 weeks, then we should still be impressed if the predictions are correct. Formalization of this leads to the notion of shattering.

Definition (Shattering) Given a set of feature vectors $\overline{x}_1, \ldots, \overline{x}_n$, we say that $\overline{x}_1, \ldots, \overline{x}_n$ are *shattered* by a class of decision rules C if all the 2^n labelings of the feature vectors $\overline{x}_1, \ldots, \overline{x}_n$ can be generated using rules from C.

Each rule from C will classify each of the feature vectors $\overline{x}_1, \ldots, \overline{x}_n$ into either class 0 or class 1. Thus, each rule splits the set of feature vectors into a subset that gets labeled 1 and its complement that gets labeled 0. There are 2^n possible subsets (or possible labelings). C shatters $\overline{x}_1, \ldots, \overline{x}_n$ if by using rules from C we can carve the feature vectors in all possible ways. This means that using a suitable rule from C, we could generate any possible prediction for the given feature vectors.

12.3 VC DIMENSION

Suppose we see labeled examples $(\overline{x}_1, y_1), \ldots, (\overline{x}_n, y_n)$. If the set of feature vectors $\overline{x}_1, \ldots, \overline{x}_n$ is shattered by C, then certainly we can find a rule from C that agrees completely with the training examples. However, from the discussion in the previous section, we expect that choosing such a rule will have little predictive power, even though it fits the data. Moreover, we will have little confidence that the rule we pick is even close to the best rule from C. We will need much more data to be confident of this.

Hence, if C shatters a large set of feature vectors, learning will be difficult if these feature vectors are observed as the training examples. The amount of data we need to learn will be large compared to the number of feature vectors shattered.

Now remember that for PAC learning, we wish to learn for *any* distributions, that is any prior probabilities $P(0)$, $P(1)$ and any conditional distributions $P(\overline{x}|0)$, $P(\overline{x}|1)$. Recall from Chapter 11 that this is equivalent to learning for any distribution $P(\overline{x})$ for the feature vector $\overline{x}$, and any conditional probabilities $P(0|\overline{x})$, $P(1|\overline{x})$. The amount of data needed for ϵ, δ learning of a class C is governed by the "bad" distributions that make learning difficult.

Thus, if C shatters *some* set of feature vectors $\overline{x}_1, \ldots, \overline{x}_n$, then for *some* distributions learning will be difficult. In particular, the distribution could be concentrated on these feature vectors, so as we see labeled examples, there are always rules agreeing with the data, but providing no information about unseen feature vectors. This discussion leads to the following definition.

Definition (VC Dimension) The *Vapnik-Chervonenkis dimension* (or *VC dimension*) of a class of decision rules C, denoted by VCdim(C), is the largest integer V such that *some* set of V feature vectors is shattered by C. If arbitrarily large sets can be shattered, then VCdim(C) = ∞.

Remember that an important point here is that we only need *some* set of V points (as opposed to *all* sets of V points) to be shattered to make the VC dimension equal to V. The VC dimension of a class is the measure of richness that we are after. As we have argued intuitively and state precisely in the next section, this measure characterizes the PAC learnability of a class C.

12.4   LEARNING RESULT

The following result is a precise statement of the fact that the VC dimension of a class of decision rules characterizes its PAC learnability. The result requires the rules in C to satisfy certain mild conditions (known as measurability conditions—see Section 3.9). These are rather technical conditions and are always satisfied in practice, so let us assume that the class C satisfies these condition.

PAC Learnability and VC Dimension
A class of decision rules C is PAC learnable if and only if the VC dimension of C is finite.

This is the characterization we were after. It shows that, as far as PAC learnability is concerned, the VC dimension of a class of rules is the "right" measure of richness.

It turns out that bounds on the amount of data needed for learning can also be obtained in terms of the VC dimension (and, of course, in terms of the parameters ϵ, δ as well).

Sample Size Upper Bound and VC Dimension

If $V = \text{VCdim}(C)$ satisfies $1 \leq V < \infty$, then a sample size

$$\frac{64}{\epsilon^2} \left(2V \log \left(\frac{12}{\epsilon} \right) + \log \left(\frac{4}{\delta} \right) \right)$$

is sufficient for ϵ, δ learning.

What is more important than the exact constants is the form of the bound and the fact that we can make such a precise statement. Of course, we should expect the sample size to grow as ϵ, δ get smaller and as V gets larger. However, the bound quantifies this in a very precise way. The behavior as a function of ϵ is $\frac{1}{\epsilon^2} \log(\frac{1}{\epsilon})$. The behavior as a function of δ is $\log(\frac{1}{\delta})$, and the behavior as a function of V is just linear (V itself).

Lower bounds can also be obtained which state that ϵ, δ learning is not possible unless a certain number of examples are used.

Sample Size Lower Bound and VC Dimension

If $V = \text{VCdim}(C)$ satisfies $1 \leq V < \infty$, then for $0 < \epsilon, \delta < 1/64$ a sample size

$$\frac{V}{320\epsilon^2}$$

is necessary. Furthermore, if C contains at least two functions, then for $0 < \epsilon < 1$ and $0 < \delta < 1/4$, a sample size

$$2 \left\lceil \frac{1 - \epsilon^2}{2\epsilon^2} \log \left(\frac{1}{8\delta(1 - 2\delta)} \right) \right\rceil$$

is necessary.

Again, the exact numbers are less important than the form of the bound and the fact that we have such a precise bound. As far as the behavior of the bounds are concerned, we can see that the lower bound is similar to the upper bound in terms of the dependence on ϵ, δ, and V.

12.5   SOME EXAMPLES

In this section we consider several examples for which computation of the VC dimension is relatively straightforward. As we see in the examples, in order to find the VC dimension of a class, the usual approach is to obtain both upper and lower bounds. If we are lucky (as we will be in the examples below), the bounds will match and we will determine the VC dimension exactly. In more complicated situations, we often have to satisfy ourselves with bounds that do not match, but give some idea of the dimension.

In order to get lower bounds on $\text{VCdim}(C)$ (that is, to show that the VC dimension is larger than some quantity), it is enough to find *some* set of feature vectors that are shattered. This is usually done by selecting some specific points and explicitly showing that all labelings (subsets) of these points can be generated by rules from C. If we find k feature vectors that are shattered by C, then we know that $\text{VCdim}(C) \geq k$.

Obtaining upper bounds (that is, showing that $\text{VCdim}(C)$ is less than some quantity) is usually more difficult. To show that $\text{VCdim}(C) < k$, we need to argue that *no* set of k feature vectors can be shattered by C. For the upper bound, it is not enough to exhibit some set of points that cannot be shattered. This usually requires a more careful argument, sometimes considering several cases for different arrangements of the feature vectors.

Example 12.1 (Intervals in 1-Dimension)   Let each feature vector consist of a single real value. Let the decision rules C be the set of all closed intervals of the form $[a,b]$ for real numbers a,b. That is, we decide 1 if $a \leq x \leq b$, and we decide 0 otherwise. What is $\text{VCdim}(C)$?

For the lower bound, it is easy to see that there are sets of two points that can be shattered. Given two points x_1 and x_2, we can find intervals containing one, or both, or neither of the points (see Figure 12.1). In fact any set consisting of two distinct points can be shattered (although recall that finding even one set is enough). Therefore, $\text{VCdim}(C) \geq 2$.

Is there *any* set of three points that can be shattered? The answer is "no," since using rules from C, we cannot label the middle point 0 and the outer points 1. That is, no interval can contain the outer points, but not contain the middle point. Therefore, we get the upper bound $\text{VCdim}(C) \leq 2$.

The upper and lower bounds together give $\text{VCdim}(C) = 2$. $\square$

Example 12.2 (Unions of Intervals)   As in the previous Example, let each feature vector consist of such a single real value. But now, let the decision rules C be the set of all finite unions of intervals. What is $\text{VCdim}(C)$?

For any k, it is easy to see that we can shatter a set of k points. Specifically, given k points, we can generate each of the 2^k labelings by putting a tiny interval

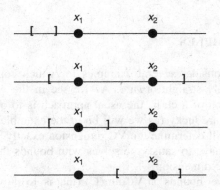

Figure 12.1   Shattering two points with intervals.

around only those points that are to be labeled 1. Since we can shatter arbitrarily large sets, we have VCdim$(C) = \infty$.

It turns out that in this example we can shatter any set of k points, even though just one set of k points for each k would be enough. □

Example 12.3 (Half-spaces in 2-D) Now, let each feature vector be a point in the plane, so that the feature space is two-dimensional. Let C be the set of all half-spaces in the plane. This is precisely the set of all decision regions with a linear decision boundary, which are the decision rules that can be represented by perceptrons. What is VCdim(C)?

For the lower bound, we can find sets of 3 points that can be shattered. In particular, take any three noncolinear points. All subsets can be generated as shown in Figure 12.2. Therefore, VCdim$(C) \geq 3$. Note that not all sets of 3 points can be shattered. Specifically, if the points are colinear then we cannot label the outer points 1 but the middle point 0.

Can any set of four points be shattered? It seems not, but how can we show this? If any three of the points are colinear, then they cannot be shattered as mentioned above. If no three of the points are colinear, then here are two generic cases to consider: (1) one point is contained in the convex hull of the other three and (2) none is in the convex hull of the others. These two cases are shown in Figure 12.3. (The convex hull of a set of points is the smallest convex set containing the set of points. So the convex hull of three points is just the triangle with the three points as the vertices.) Note that the first case also includes arrangements where three or all the four points are colinear.

In the first case, we cannot generate the labeling where the inner point (the one in the convex hull) is 0, while the others are labeled 1. Any half-space containing

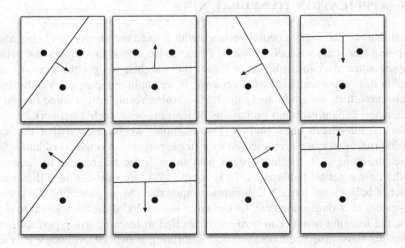

Figure 12.2 Shattering three points with half-spaces.

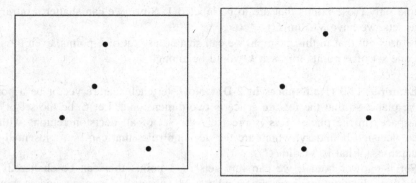

Figure 12.3 Two generic arrangements of four points.

the outer three points must also contain the fourth (inner) point. Therefore, no set of four points in this case can be shattered.

In the second case, we cannot generate the labeling where two opposite corners are labeled 1 while the other two opposite corners are labeled 0. The reason is that since half-spaces are convex, if two points are labeled 1, then all points on the line joining these two points must also be labeled 1. However, this is also true of two points labeled 0. But this means that the intersection of the two line segments must be labeled both 0 and 1, which is a contradiction. Therefore, no set of four points can be shattered.

The result is that there is no set of four points that can be shattered. Therefore, VCdim(C) $\leq$ 3, which combined with the lower bound gives VCdim(C) = 3. $\square$

12.6 APPLICATION TO NEURAL NETS

Recall that we can view a neural network with a fixed architecture as being able to compute some class of decision rules. Each choice of weights makes the network compute some decision rule. As we vary the weights, we get the class C of all decision rules representable by the network. If we could compute the VC dimension of this class, then we could apply the PAC learning results to get some insight into the inherent capabilities and limitations of learning with such a network.

Such results have been studied. For example, we have seen that the set of decision rules that can be represented by a single perceptron is the set of half-spaces. If the threshold is 0, the half-spaces pass through the origin, but for adjustable thresholds we get all half-spaces. In Example 12.5, we saw that in 2 dimensions, the set of half-spaces has a VC dimension equal to 3. More generally, the set of all half-spaces in d dimensions can be shown to have a VC dimension equal to $d + 1$. Thus, the learning results can be directly applied to learning using perceptrons.

Results have also been obtained for multilayer networks. Although it is quite difficult to compute the VC dimension exactly, useful bounds have been obtained.

For networks with threshold units, one bound is that

$$\text{VCdim}(C) \le 2(d + 1)(s) \log(es),$$

where d is the dimension of the feature space, s is the total number of perceptrons, and e is the base of the natural logarithms (approximately 2.718).

Similar results have been obtained for the case of sigmoidal units. In this case, the bound also involves the maximum slope of the sigmoidal output function and the maximum allowed magnitude for the weights.

Such bounds allow the PAC learning results to be applied, giving a theoretical justification for the required sample size for training a neural network.

12.7 SUMMARY

In this chapter, we discussed the trade-offs involved in learning in terms of the richness of the class of decision rules being used. A very rich class C is useful to reduce the approximation error, which refers to the inability of rules from C to approximate the Bayes decision rule. On the other hand, if C is very rich, then using only the training examples, we will have a hard time identifying which rule from C will perform the best on future examples, which is referred to as the estimation error. To quantify the notion of "richness" of a class of decision rules, we introduced the notion of "VC dimension" and the notion of "shattering" it appeals to. VC dimension is one way to measure the "expressive power" of the class of rules C which is more important in learning than just the number of rules in C. The more the data points that can be shattered by hypotheses in the class C, the more difficult PAC learning is for that class. In fact, a class C is PAC learnable if and only if the VC dimension of C is finite, and moreover the amount of data needed for ϵ, δ learning depends on the VC dimension. In the next chapter, we discuss learning when the class of rules C has infinite VC dim.

12.8 APPENDIX: VC DIMENSION AND POPPER DIMENSION

The role of VC dimension in statistical learning theory is reminiscent of the emphasis on "falsifiability" in Karl Popper's philosophy of science. Popper argued that evidence cannot establish a scientific hypothesis; it can only "falsify" it. A scientific hypothesis is a falsifiable *conjecture*. A useful scientific hypothesis is a falsifiable conjecture that has withstood empirical testing.

Recall that PAC learning requires an initial choice of a set of rules C. In Popper's terminology, such a choice involves the "conjecture" that the relevant rules are the rules in C. According to Popper, if this conjecture is to count as scientific rather than "metaphysical," the class of rules C must be appropriately "falsifiable."

Popper argues that there are degrees of difficulty of falsifiability. For example, he argues that a linear hypothesis is more falsifiable—easier to falsify—than a quadratic hypothesis. This seems to fit with the role of VC dimension, because the collection of linear classification rules has a lower VC dimension than the collection of quadratic classification rules.

However, Popper's measure of the degree of falsifiability of a class of hypotheses does not quite correspond to VC dimension. Where the VC dimension of a class C of hypotheses is the largest number N such that *some* set of N points is shattered by rules in C, what we might call the "Popper dimension" of the difficulty of falsifiability of a class is the largest number N such that *every* set of N points is shattered by rules in C. This difference between *some* and *every* is very important and VC dimension turns out to be the key notion in statistical learning theory, rather than Popper-dimension.

For example, the class of linear hypotheses in d-space has a VC dimension of $d + 1$. *Some* set of $d + 1$ points can be shattered. But that class has a Popper dimension of 2, because three colinear points cannot be shattered. So, if $m > 2$, not *all* points can be shattered.

This suggests that Popper's theory of falsifiability might be improved by adopting VC dimension as the relevant measure in place of his own measure.

12.9 QUESTIONS

1. True or False: A set of k points is shattered by a class of rules C if all 2^k labelings of the points can be generated using rules from C.

2. True or False: VCdim(C) is the largest integer v such that every set of v points can be shattered by C.

3. True or False: VCdim(C) is one plus the number of parameters needed to specify a particular rule from the class C.

4. If VCdim(C) = v, what is the smallest number of rules that the class C could contain?

5. If C is a class of decision rules, what condition on VCdim(C) is needed for PAC learnability?

6. If C is a class of decision rules, what are R^*, $\hat{R}(h)$, and R_C^*?

7. What relations of the form $X \leq Y$ hold between R^*, $\hat{R}(h)$, and R_C^*?

8. Is the class of rules representable by a perceptron PAC learnable?

9. Consider the class C of all orthogonal rectangles in the plane, that is, all rectangles whose sides are parallel to the coordinate axes. What is VCdim(C)? Justify your answer.

10. Consider the class C of all convex subsets of the plane. What is VCdim(C)? Justify your answer. (Hint: think about points on the circumference of a circle.)

11. In PAC learning, discuss the advantages and disadvantages of requiring that the same sample size $m(\epsilon, \delta)$ work for *every* choice of prior probabilities and conditional densities. How might we modify the definition of PAC learning model to relax this requirement?

12. (a) In PAC learning, suppose we let the learner choose the feature points he/she would like classified instead of providing randomly drawn examples. Suggest a way in which you might measure how much "learning" has taken place in this case, that is, what sort of performance criterion might you use to measure success after the learner has asked for some number of examples?

   (b) For your answer to part (a), does this make learning easier or harder than standard PAC learning?

13. If the correct classification of an item is completely determined by its observable features, what is the Bayes error rate for decision rules using those features? In this case, does it follow that C is PAC learnable?

14. One might argue that every learning algorithm works with a fixed set of decision rules, namely, the set of all rules that the particular algorithm might possibly produce over all possible observations. In light of such an argument, what are the advantages of explicitly specifying a class of rules C and studying the problem of learning using rules from this class?

15. If one learning algorithm works with a class of rules that includes all the rules of a different algorithm and some others, does this mean that the former algorithm should give a strictly better performance in a learning task? Why or why not? Discuss any advantages/disadvantages to be had by restricting the set of decision rules.

12.10 REFERENCES

The notions of estimation and approximation errors are widely used in statistics. VC dimension is one very useful way of quantifying these ideas. Books or review articles that discuss PAC learning invariably discuss VC dimension as well, including the references Kearns and Vazirani (1994), Devroye *et al.* (1996), Vapnik (1996), Vidyasagar (1997), Kulkarni *et al.* (1998), and Anthony and Bartlett (1999) mentioned in the previous chapter.

The notion of VC dimension first appeared in (and is named after) the work of Vapnik and Chervonenkis (1971). A great deal of further work on this idea was done in the probability, statistics, and pattern recognition communities. Blumer *et al.* (1989) showed the connection between VC dimension and the model of learnability proposed in the computer science community by Valiant (1984). This led to renewed interest in PAC learning and related topics and a convergence of work in the various communities.

Corfield *et al*. (2009) discuss the relation between Popper's (1979, 2002) measure of falsifiability and VC dimension. Our discussion here mostly repeats Harman and Kulkarni (2007), pp. 50–52, deriving from conversations with Vladimir Vapnik.

Anthony M, Bartlett PL. Neural network learning: theoretical foundations. Cambridge: Cambridge University Press; 1999.

Blumer A, Ehrenfeucht A, Haussler D, Warmuth M. Learnability and the Vapnik-Chervonenkis dimension. J ACM 1989;36(4):929–965.

Corfield D, Schölkopf B, Vapnik V. Falsificationism and statistical learning theory: comparing the popper and Vapnik-Chervonenkis dimensions. J Gen Philos Sci 2009; 40:51–58.

Devroye L, Györfi L, Lugosi G. A probabilistic theory of pattern recognition. New York: Springer-Verlag; 1996.

Harman G, Kulkarni S. Reliable reasoning: induction and statistical learning theory. Cambridge (MA): MIT Press; 2007.

Kearns MJ, Vazirani UV. An introduction to computational learning theory. Cambridge (MA): MIT Press; 1994.

Kulkarni SR, Lugosi G, Venkatesh S. Learning pattern classification—A survey. IEEE Trans Inf Theory 1998;44(6):2178–2206.

Mitchell TM. Machine learning. New York: McGraw-Hill; 1997.

Popper K. Objective knowledge: an evolutionary approach. Oxford: Clarendon Press; 1979.

Popper K. The logic of scientific discovery. London: Routledge; 2002.

Valiant LG. A theory of the learnable. Commun ACM 1984;27(11):1134–1142.

Vapnik VN. The nature of statistical learning theory. New York: Springer-Verlag; 1996.

Vapnik VN, Chervonenkis A. On the uniform convergence of relative frequencies of events to their probabilities. Theory Probab Appl 1971;16(2):264–280.

Vidyasagar M. A theory of learning and generalization. London: Springer-Verlag; 1997.

CHAPTER 13

Infinite VC Dimension

In the previous chapter, we saw that for a fixed class of decision rules C, the VC dimension of C characterizes whether or not PAC learnability is possible. Under mild regularity assumptions, C is PAC learnable if and only if the VC dimension of C is finite. In this case, for a given accuracy and confidence, some finite number of examples are sufficient to choose a near-optimal rule from C with high probability.

As mentioned in the previous chapter, this type of result focuses on the *estimation* error, the problem of trying to predict which rule from C will be the best on the basis of a set of random examples. If the class of rules is too rich, then even with a large number of examples, it can be difficult to distinguish those rules from C that will actually perform well on new data from those that just happen to fit the data seen so far but have no predictive value.

In addition to the estimation error, the *approximation* error is also important in determining the overall performance of a decision rule. The approximation error sets a limit on how well we can do, no matter how much data we receive. If we are restricted to using rules from the class C, we can certainly do no better than the best rule from C. Once we fix a class with finite VC dimension, we are stuck with whatever approximation error results from the class and the underlying distributions. Moreover, since the distributions are unknown, the actual approximation error we must live with is also unknown.

In the notation we have been using, the best we can ever hope for is to select a rule with the Bayes error rate R^*. But using rules from C, the best we can hope for is R_C^*. Sometimes R_C^* might be significantly larger than R^*. What if we are not satisfied with this? Is it unreasonable to want a rule with an error rate close to R^* if we have a large amount of training data?

If we think about the nearest neighbor and kernel rules, we see that it is reasonable to expect to do as well as possible (Bayes error rate) as we get more and more data, although the criterion there was not exactly the PAC criterion. We know from the previous chapter that the PAC criterion is too stringent since we are restricted

An Elementary Introduction to Statistical Learning Theory, First Edition.
Sanjeev Kulkarni and Gilbert Harman.
© 2011 John Wiley & Sons, Inc. Published 2011 by John Wiley & Sons, Inc.

to classes with finite VC dimension. But if we slightly modify the PAC criterion, then we can deal with certain classes that have infinite VC dimension, and thereby address the issue of nonzero approximation errors.

13.1 A HIERARCHY OF CLASSES AND MODIFIED PAC CRITERION

Let us consider a sequence of classes $C_1, C_2, \ldots$ that are nested, so that $C_1 \subset C_2 \subset \cdots$. We insist that each of the C_i has finite VC dimension, but let the VC dimension grow unboundedly with i. That is, if $V_i = \text{VCdim}(C_i)$, then we require $V_i < \infty$ for all i, but we allow $V_i \to \infty$ as $i \to \infty$.

Given such a sequence of classes, we can think of $\text{VCdim}(C_i)$ as a measure of the "complexity" of the *class* of rules C_i. We can identify the complexity of a *rule* with the smallest index of a class C_i to which the rule belongs. The "simplest" decision rules (by this criterion) are in C_1; the rules in C_2 include the simplest rules as well as some slightly more complex rules; and so on.[1] The collection of all rules under consideration is given by

$$C = \bigcup_{i=1}^{\infty} C_i.$$

If $V_i \to \infty$, then the class C must have infinite VC dimension since it has a dimension at least as large as V_i for each i. Hence, the class C is too rich (as measured by VC dimension) to be PAC learnable, although each of the C_i on its own is PAC learnable.

Nevertheless, it may turn out that we can still find good rules from C as we get more and more data. The criterion for finding a good rule cannot be in the PAC sense, since we already know that C is not PAC learnable. We need to relax the PAC criterion in a way that is still useful but general enough to handle the nested structure of decision rules. The key idea is to allow the number of examples needed for ϵ, δ learning to depend on the underlying probability distributions (as well as on ϵ and δ), which is both an intuitive and reasonable idea. We should expect that complicated or difficult problems will require more data for learning, while simple problems can be learned easily with minimal data.

13.2 MISFIT VERSUS COMPLEXITY TRADE-OFF

With this modified PAC criterion, a hierarchy of classes $C = \cup_{i=1}^{\infty} C_i$ can be learned by trading off the fit to the data against the complexity of the hypothesis. As above, by "complexity" we mean the smallest index of the class to which the hypothesis belongs.

---

[1] In Chapter 16 we note ways in which this way of thinking about simplicity may contrast with more ordinary notions.

This idea is at the heart of various techniques that go by different names: Occam's razor, minimum description length (MDL) principle, structural risk minimization, Akaike's information criterion (AIC), etc. The intuition behind these approaches is the following. If we have only a small amount of data, then we should entertain only simple hypotheses. The data simply do not contain enough information/evidence to draw highly complex conclusions. However, as we get more and more data, we should be willing to entertain increasingly complex hypotheses. But among the hypotheses under consideration, which one should we choose? It makes sense to favor those that fit the data well. This is what we did in the standard PAC case, where in fact, we simply chose the hypothesis that fit the data the best. Now, we have the extra insight that it is not only the fit to the data that matters but the complexity of the hypothesis is also a consideration. We should favor hypotheses of lower complexity and/or let the complexity of hypotheses we consider grow as we get more data.

Thus, our final choice of hypothesis should reflect some trade-off between misfit on the data (error) and complexity. Given specific ways of measuring the error of a decision rule on the data and the complexity of a decision rule, we can select a hypothesis by

$$h = \operatorname{argmin}_{h \in C} \; \text{error}(h) + \text{complexity}(h).$$

The notation $\operatorname{argmin}_{h \in C}$ means that we should find the hypothesis in C that minimizes the given expression. Making the first term small favors choosing a hypothesis in some C_i with large i. These are the richer classes and can fit the data better. Presumably, if i is large enough, then we would be able to fit the data exactly. However, this would result in "overfitting" and would not provide much predictive power.

The second term helps to control this "overfitting" problem. The second term is small for hypotheses in C_i for small i. This could be made quite small (say equal to 1), by simply selecting some hypothesis from C_1. However, this class of decision rules presumably has very restrictive representational capabilities, and rules from this class may not be able to capture the structure in the data. A hypothesis from this class will likely make many errors on the training data, and hence also may not serve so well in prediction.

By striking a balance between these two terms, we control the expressive power of the rules we will entertain, and also consider how well the rules take into account the training data. With specific choices for this type of trade-off, the class C is learnable in the modified PAC sense.

13.3 LEARNING RESULTS

There are many choices for exactly how to carry out the misfit versus complexity trade-off. One way to measure the misfit of a hypothesis on the data is to just count the number of errors, as we did before. One way to measure the complexity of the hypothesis is (as suggested above) to take the index of the first class to which it belongs. But many variations are possible. For example, we can multiply either term by a constant (i.e., take the complexity as before plus twice the number of errors). We could take the square of the complexity (as a new measure of complexity). We could choose a complexity parameter k_n based on the number of examples n that we have seen, and then find the best hypothesis from the class C_{k_n}.

One concrete result is as follows. As before, let V_i denote the VC dimension of C_i, where $V_i < \infty$ for all i but $V_i \to \infty$ as $i \to \infty$. Further, suppose that $R^*_{C_i} \to R^*$ as $i \to \infty$. This means that we can find rules with performance arbitrarily close to the optimal Bayes decision rule as i gets sufficiently large. We select a sequence $k_n \to \infty$ such that

$$\frac{V_{k_n} \log(n)}{n} \to 0 \text{ as } n \to \infty. \tag{13.1}$$

After seeing n examples, we select a rule h_n from the class C_{k_n} that fits the data the best. Then it can be shown that this method is universally consistent. That is, for any distributions, as $n \to \infty$, we have $R(h_n) \to R^*$. Hence, as we get more and more data, the performance of the rule we select gets closer and closer to the Bayes error rate.

13.4 INDUCTIVE BIAS AND SIMPLICITY

There may also be many choices for the hierarchical breakdown of C into the constituent C_i as well. Any choice can be made to work as long as each C_i has finite VC dimension and the union of the C_i contains all the rules in C. Often there is a natural decomposition based on the description of rules in C. The complexity decomposition used in learning need not conform to this natural decomposition, but often it will by choice.

The various choices for measuring error on the data, the complexity of decision rules, and the misfit versus complexity trade-off reflect different types of inductive bias. The choices dictate exactly how the learner strikes a balance between preferring certain hypotheses (those of low complexity, or equivalently, simple ones), and how much weight is to be given to the evidence in the data.

As mentioned, many different choices can be made to work in the modified PAC sense if carried out appropriately. That is to say, as we get more data, we can find better hypotheses, and under certain conditions we can guarantee that the error rate of our hypotheses converges to the Bayes error rate as the number of examples $n \to \infty$. However, we cannot, in general, obtain uniform sample size bounds for which we will be guaranteed to have ϵ, δ learning after some fixed number of examples. The number of examples needed will depend on the underlying distributions, and these are unknown. The actual performance in practice will depend on how well our inductive bias reflects the learning problems we encounter. In engineering applications, these choices are often a matter of art, intuition, and technical understanding of the problem domain. One of the key choices is the complexity hierarchy of the decision rules, which reflects the learner's bias on which rules are considered "simple". Simplicity as a tool in inductive reasoning has been very broadly used and is discussed further in Chapter 16.

13.5 SUMMARY

We noted that the higher the VC dimension of the hypothesis set C, the more likely it is that the best hypothesis in C approximates the Bayes error rate but also the longer it takes to attain PAC learning. This reflects a trade-off between estimation error and approximation error, but rather harshly forces large approximation by requiring finite VC dimension. By relaxing the PAC condition to allow the number of examples to depend on the underlying distribution, we can deal with classes that have infinite VC dimension by decomposing the class into a nested infinite union of classes C_i where each C_i has finite VC dimension. We think of the C_i as a decomposition based on the complexity of the decision rules. After seeing some labeled examples, the learner selects a hypothesis by suitably trading off the misfit to the data and the complexity of the hypothesis. We get a type of modified PAC learning in the limit with appropriate choices for the misfit/complexity balance. One of the key choices is the way we measure the complexity, or equivalently, the simplicity, of the decision rules. This approach is related to a number of other learning methods based on simplicity known by various names such as Occam's razor, MDL principle, structural risk minimization, AIC, etc.

13.6 APPENDIX: UNIFORM CONVERGENCE AND UNIVERSAL CONSISTENCY

It is important to distinguish *uniform convergence* from *universal consistency*. These expressions sound alike but indicate different desirable properties of a learning problem.

The results about PAC learning described in Chapter 11 are results about uniform convergence. Where C has a finite VC dimension, there is a function $m(\epsilon, \delta)$ that provides an upper bound to the amount of data needed to guarantee a certain probability $(1 - \delta)$ of endorsing rules whose expected error is within ϵ of the minimum expected error for rules in C. This is a uniform convergence result in that the amount of data needed for a given level of performance (as specified by ϵ and δ) does not depend on the underlying distributions characterizing the problem. The same $m(\epsilon, \delta)$ works for any distributions (and hence no matter which rule from C is the best). However, the minimum expected error for rules in C may not be close to the expected error of a best rule, a Bayes rule.

In this chapter we discussed an extension of the setup from Chapter 11 for which there are learning methods that are universally consistent in the sense that for any background probability distribution, with probability approaching 1, as more and more data are obtained, the expected error of rules endorsed by the method approaches in the limit the expected error of a Bayes rule. The nearest neighbor and kernel rules that we discussed in previous chapters are also universally consistent under appropriate constraints.

Universal consistency does not imply uniform convergence. There may be no bound on the amount of data needed in order to ensure that (with probability approaching 1) the expected error of the rules endorsed by the method will be within ϵ of the expected error of a Bayes rule. Nevertheless, universal consistency is clearly a desirable characteristic of a method. It does provide a convergence result, because the error rate of the rule endorsed by a universally consistent method converges to the expected error of a Bayes rule for any background probability distribution. Although this does not guarantee a rate of convergence, it can be shown that no method provides such a guarantee for all distributions. However, given certain assumptions on the background distribution, the uniform convergence rates can be obtained.

13.7 QUESTIONS

1. Consider the class C discussed above which is the union of $C_1, C_2, \ldots$. Do the various C_i have finite VC dimension? Does C? Explain.

2. Suppose a class of decision rules C has infinite VC dimension. Is it ever possible to decompose C into a hierarchy $C_1 \subset C_2 \subset \cdots$ such that $C = \cup_{i=1}^{\infty} C_i$ but with $\lim_{i \to \infty} V_i < \infty$? Why or why not?

3. In the learning method described in Section 13.3, show that for any choice of the C_i with $V_i \to \infty$ there is a choice of k_n which violates Equation (13.1). What problem might the hypotheses h_n encounter in this case?

4. In Equation (13.1), why does the choice of k_n need to depend on the VC dimension of C_{k_n} instead of a condition like $k_n/n \to 0$ or $k_n \log(n)/n \to 0$?

5. True or False: With a suitable choice of the class of rules $\mathcal{C}$, it is possible to have both universal consistency and uniform convergence.

6. Is PAC learning a matter of uniform convergence, universal consistency, or both?

7. Suppose we have a nested collection of sets of decision rules $\mathcal{C}_1 \subset \mathcal{C}_2 \subset \mathcal{C}_3 \subset \cdots$ each with finite VC-dimension $d_i = \text{VCdim}(\mathcal{C}_i) \to \infty$. After n examples, we select a decision rule h_n from the class $\mathcal{C}_{k_n}$ that fits the data the best. That is, after n examples, we consider rules of "complexity" k_n and among these we use empirical risk minimization.

   (a) What happens if k_n is some fixed constant k that does not grow with n? Does the estimation error go to 0? Does the approximation error go to 0?

(b) What happens if $k_n \to \infty$ extremely fast compared with d_n? Does the estimation error go to 0? Does the approximation error go to 0?

(c) What hypothesis do we select if we ignore the complexity of the hypothesis?

(d) What hypothesis do we select if we ignore the error on the data?

8. When scientists accept an inverse square law of gravitation rather than some more complicated law that would account equally well for the data, is there a way to justify their conclusion without supposing that nature is simple? Explain. How does this relate to the ideas discussed in this chapter?

13.8 REFERENCES

The appeal to technical criteria of simplicity in statistics and learning is widespread and has a long history. Its use in the context of VC theory was proposed by Vapnik (1982, 1996), using the name *structural risk minimization*. See also, for example, Chapters 2 and 3 of Kearns and Vazirani (1994), Section V of Kulkarni *et al.* (1998), Chapter 18 of Devroye *et al.* (1996), and references therein. The references use the terms Occam's razor and complexity regularization to refer to the approach. Occam's razor is a broadly used term for techniques that make use of simplicity, based on a principle of the fourteenth century philosopher William of Occam (or Ockham) that states "Entities should not be multiplied without necessity." Other techniques in statistics and learning related to Occam's razor include Rissanen's (1989) MDL principle and Akaike's (1974) AIC. Some work in the philosophy literature that discusses these ideas in the context of model selection includes Sober (1990) and Zellner (1999).

Akaike H. A new look at the statistical model identification. IEEE Trans Automat Control 1974;19:716–723.

Devroye L, Györfi L, Lugosi G. A probabilistic theory of pattern recognition. New York: Springer-Verlag; 1996.

Kearns MJ, Vazirani UV. An introduction to computational learning theory. Cambridge (MA): MIT Press; 1994.

Kulkarni SR, Lugosi G, Venkatesh S. Learning pattern classification—A survey. IEEE Trans Inf Theory 1998;44(6):2178–2206.

Rissanen J. Volume 15, Stochastic complexity in statistical inquiry, Singapore: World Scientific, Series in Computer Science; 1989.

Sober E. Let's Razor Occam's Razor. In: Knowles K, editor. Explanation and its limits. Cambridge: Cambridge University Press; 1990.

Vapnik VN. Estimation of dependencies based on empirical data. New York: Springer-Verlag; 1982.

Vapnik VN. The nature of statistical learning theory. New York: Springer-Verlag; 1996.

Zellner A, Keuzenkamp H, McAleer M,editors. Simplicity, inference and econometric modelling. Cambridge: Cambridge University Press; 2002.

CHAPTER 14

The Function Estimation Problem

So far, we have been discussing learning in connection with pattern classification. Now we want to consider a different formulation involving estimation rather than classification. That is, instead of deciding between two classes (0 and 1), we wish to estimate a real value. In this chapter, we discuss the basic formulation of the estimation problem. In Chapter 15 we discuss some learning methods and results for estimation analogous to the ones we have discussed so far for classification.

14.1 ESTIMATION

In the standard pattern classification problem that we discussed in Chapter 4, we observe a feature vector $\overline{x}$ and we want to guess the associated label y. In the case of classification that we have been discussing, the label has taken on one of two values ($\{0, 1\}$ or $\{-1, 1\}$). A finite set of values for the label y can also be considered, though we do not address that case in this book.

In the estimation (or regression) problem, the "label" y can take on an arbitrary real value and our estimate $f(\overline{x})$ will also be a real value. Now, instead of decision rules being partitions, we have functions $f(\overline{x})$. For each feature vector $\overline{x}$, the estimate $f(\overline{x})$ is a real number, so the estimator is a mapping from $\mathbf{R}^d$ to $\mathbf{R}$.

As before, we assume that there are background probability distributions that govern how $\overline{x}$ and y are drawn and related. These can be described by a joint probability distribution $P(\overline{x}, y)$ or a joint density $p(\overline{x}, y)$ in the $(d + 1)$-dimensional space $\mathbf{R}^d \times \mathbf{R}$.

In the case of estimation, it is often convenient to think of this joint distribution in terms of a probability distribution for the feature vector (or independent variable) $\overline{x}$ and a conditional probability distribution for the label (or dependent variable) y.

An Elementary Introduction to Statistical Learning Theory, First Edition.
Sanjeev Kulkarni and Gilbert Harman.
© 2011 John Wiley & Sons, Inc. Published 2011 by John Wiley & Sons, Inc.

That is, $p(\overline{x}, y)$ is determined by the distribution (density) $p(\overline{x})$ from which the feature vector $\overline{x}$ is drawn and the conditional distribution (density) $p(y|\overline{x})$ from which y is drawn, given $\overline{x}$.

14.2 SUCCESS CRITERION

The fact that the label y is now a real value requires us to reconsider the criterion for success. In the case of classification, our success criterion is the probability of error, that is, the probability that our decision does not match the true label y. But as soon as y can take on a continuum of values, trying to estimate y exactly is unreasonable. The probability of error needs to be replaced by some other performance criterion.

A natural approach is to introduce a loss function (or cost function) to measure how close our estimate $f(\overline{x})$ is to the observed value y. We might try to measure loss by the difference $y - f(\overline{x})$. The problem with this measure is that we are often interested in the average loss. By taking the difference, if we sometimes grossly overestimate y and other times we grossly underestimate y, these large negative and positive losses can cancel out. This can lead to zero (or close to zero) average loss even when the loss for any given $\overline{x}$ is huge.

To remedy this, we might measure loss by absolute error $|y - f(\overline{x})|$. In this case, the loss is always positive, so that huge losses for different values of $\overline{x}$ cannot cancel each other out. Another way to avoid this problem is to consider squared error loss given by $(y - f(\overline{x}))^2$. The squared error is often easier to work with than the absolute loss because it has a smooth derivative so that machinery from calculus can be more readily used to analyze it. We saw this in the discussion of backpropagation for multilayer perceptrons as well (Chapter 10). It is also possible to use other loss functions, but squared error loss is the most common and is what we use.

A best estimator is one that minimizes average (or expected) loss. We now discuss the average loss, and discuss in the next section the estimator that minimizes this loss. Suppose we have an estimator $f(\cdot)$, and let us fix a particular feature vector $\overline{x}$ for now. In this case, our estimate is $f(\overline{x})$ so that if a particular value of y is observed, then the loss is $(y - f(\overline{x}))^2$. But y is drawn randomly from the conditional distribution $p(y|\overline{x})$. Hence, given this fixed $\overline{x}$, the expected loss, denoted by $R(f, \overline{x})$, is given by

$$R(f, \overline{x}) = \int_{\mathbf{R}} (y - f(\overline{x}))^2 \, p(y|x) \, dy \tag{14.1}$$

But recall that $\overline{x}$ is also drawn randomly, namely according to the distribution $p(\overline{x})$. Hence, the expected loss of the estimator $f(\cdot)$ averaged over all values of $\overline{x}$ is given by

$$R(f) = \int_{\mathbf{R}^d} R(f, \overline{x}) \, p(\overline{x}) \, d\overline{x} = \int_{\mathbf{R}^d} \int_{\mathbf{R}} (y - f(\overline{x}))^2 \, p(y|\overline{x}) \, p(\overline{x}) \, dy \, d\overline{x}. \tag{14.2}$$

14.3   BEST ESTIMATOR: REGRESSION FUNCTION

In classification, a best estimator (Bayes Rule) minimized the probability of error. Equivalently, it minimizes the average loss. If the loss is 1 if we make an error and is 0 if we decide correctly, then the probability of error is precisely the average loss.

Likewise, for estimation, we seek an estimator (or function) that minimizes the average loss in Equation (14.2). To find such an estimator, note that the average loss R is the average of the conditional loss $R(\overline{x})$ over the distribution $p(\overline{x})$. So, to minimize R, we need to minimize $R(\overline{x})$ for every $\overline{x}$.

Consider a number a that is to be drawn at random, and suppose we want to estimate a by some fixed number b. What is the average squared error of this estimate and what number b would give us the smallest average squared error? Since b is fixed, we can see that

$$E[(a - b)^2] = E[a^2 - 2ab + b^2]$$
$$= E[a^2] - 2E[a]b + b^2,$$

where $E[a]$ is just the average value of a.

The first term does not depend on our estimate b. So, to find which number b minimizes the average squared error, we need only to find which b minimizes $b^2 - 2E[a]b$. The value of b that does this is just the expected value (or average, or mean) of a, namely, $b = E[a]$.

For our original estimation problem, we can basically just do this, but for every $\overline{x}$. Consider a given value of $\overline{x}$. The label y is random and comes from the distribution $p(y|\overline{x})$. As we discussed, for squared error loss, the best estimate for y is the average of y (given this $\overline{x}$), namely the conditional mean of y, given $\overline{x}$.

For a given $\overline{x}$, let $m(\overline{x})$ denote this expected value of y, given $\overline{x}$. This number $m(\overline{x})$ is the estimator for which $R(\overline{x})$ is minimal. Thus, as a function of $\overline{x}$, $m(\overline{x})$ is precisely the estimator we are looking for; that is, the one that minimizes the average squared loss. This function is often called the *regression function*.

The corresponding minimum average loss (Bayes loss or Bayes error rate), denoted by R^* as before, is the expected loss when the regression function is used as the estimator. This is analogous to the case of classification for which the best decision rule (Bayes rule) is a partition that minimizes the probability of error. For estimation, the best rule (the regression function) is a function that minimizes the average loss.

14.4   LEARNING IN FUNCTION ESTIMATION

In the case of classification, if we knew the underlying distributions, we could compute Bayes rule. In the case of estimation, if we knew the underlying distributions, we could directly find the regression function $m(\overline{x})$ by formally computing the conditional mean of y given $\overline{x}$.

However, as before, we generally know little or nothing about the underlying distribution $p(\overline{x}, y)$. Instead, assume that we have labeled data assumed to be produced by the same underlying probability distribution—independent and identically distributed (i.i.d.) examples drawn according to $p(\overline{x}, y)$. This training data are denoted by $(\overline{x}_1, y_1), (\overline{x}_2, y_2), \ldots, (\overline{x}_n, y_n)$ as before.

Given the training data, we wish to find a function that is close to the regression function $m(\overline{x})$ in the sense of average squared error. Our benchmark (the best we can possibly achieve) is the Bayes rate R^*, which is the squared error obtained when using the regression function $m(\overline{x})$ as the estimate of the label y.

14.5 SUMMARY

In this chapter we discussed a probabilistic formulation for the estimation problem. As in the classification problem, we assume that there is a prior probability distribution $p(\overline{x}, y)$ characterizing the problem, where $\overline{x}$ is the feature vector and y is the label. The fundamental difference between classification and estimation is the range of values that y may take. For classification, y takes on one of two values, say 0 or 1. While we have not discussed it here, the case of a finite number of classes is in many ways similar to the two-class problem. However, in the estimation problem, the label y can be a real number.

This basic difference requires us to modify our success criterion. Rather than considering the probability of error, we seek decision rules that minimize the average squared error. The best estimator, the Bayes rule, is given by the conditional average of y given $\overline{x}$, which we denoted by $m(\overline{x})$. This function is also called the regression function. The learning problem in the case of estimation becomes one of estimating the regression function $m(\overline{x})$ on the basis of training data $(\overline{x}_1, y_1), \ldots, (\overline{x}_n, y_n)$ that is drawn from $p(\overline{x}, y)$. In the next chapter, we discuss some methods and results for estimation, which are extensions of some of the methods we have discussed for classification.

14.6 APPENDIX: REGRESSION TOWARD THE MEAN

Francis Galton used the term "regression" to name an interesting biological phenomenon, and this term was later used in a broader statistical context. The heights of adult children of very tall parents tend on average to be less than the heights of their parents. These heights tend to "regress" toward average or mean heights. This sort of regression toward the mean is a statistical phenomenon, because some adult children of very tall parents are even taller than their parents, but on average the adult children of tall parents are less tall. On the other hand, the heights of adult children of very short parents are on average less than the heights of their parents. Here too there is regression in the direction of average height.

Here is an oversimplified explanation of this statistical phenomenon. Suppose there are factors that affect height and that these factors vary from one population to another and are passed on from parent to child. The factors for a height of h have a probabilistic relation to adult height, which might be represented by a bell-shaped probability density curve whose highest point is h. If the density is symmetric, then the probability that people with these factors have an adult height of $h + 1$ will be the same as the probability that they have an adult height of $h - 1$. Assume that the distribution of height factors of people in this population can also be represented by a bell-shaped curve whose highest point is p. This point represents the 50th "percentile" of the distribution of the height factors, half the population having lower height factors and half having higher height factors. It also almost certainly represents the 50th percentile of adult heights in the population.

Suppose that p is 5' 6'' and consider those people whose adult height h is 6' 6''. Some of those people have inherited height factors from a parent whose adult height is exactly h, some have inherited height factors from a parent whose adult height is slightly greater than h, and some have inherited height factors from a parent whose adult height is slightly less than h. Because the heights and height factors of adults in the population form a bell-shaped curve centered on p, there are more people in the population whose adult heights and height factors are slightly less than h than people whose adult heights and height genes are slightly more. So, people whose adult height is h are more likely to have height factors of somewhat less than h, so the expected heights of the adult children of such people are likely to be less than h.

14.7 QUESTIONS

1. If a loss function other than squared error is used, will the regression function (i.e., the conditional mean of y, given $\bar{x}$) still always be the best estimator?

2. Discuss the quip "Even a stopped clock is right twice a day," and contrast it with the addition that "and a working clock is almost never exactly right."

3. (a) Suppose we consider the function estimation problem (rather than the classification problem), but insist on using the probability of error as our success criterion (rather than squared error). That is, we "make an error" whenever our estimate $f(x)$ is not equal to y. What is the smallest probability of error we could hope for in the general case with densities?

   (b) Now suppose we consider squared error, but the distributions are such that the y value is always equal to either 0 or 1. If $P(y = 1|x) = p$ and $P(y = 0|x) = 1 - p$, what estimate $f(x)$ minimizes the squared error?

4. True or False: The Bayes error rate for estimation problems (as opposed to classification problems) can be greater than 1/2.

5. True or False: In the function estimation problem, if x is the observed feature vector and y is a real value that you wish to estimate, the optimal Bayes

decision rule (for least squared error) dictates that we estimate y according to the maximum posterior probability, that is, find the maximum of $p(y|x)$.

14.8 REFERENCES

Estimation in statistics refers to a broad class of methods used for estimating parameters, functions, or other quantities on the basis of observations that usually contain some randomness. There are many books on statistics that deal with various aspects of estimation.

The estimation problem considered here is often referred to as regression estimation, function estimation, or sometimes curve estimation. The problem of regression estimation goes back to the early 1800s to the work of Legendre and Gauss. Much of the work on regression estimation deals with the parametric case in which the form of the underlying function is known and only certain parameters need to be estimated. As with the case of classification, our interest is the case of nonparametric estimation in which very little or nothing is assumed or known about the underlying function. There are a number of books and papers on nonparametric estimation. Some of the books are Corder and Foreman (2009), Györfi, *et al*. (2002), and Hardle (1992). The paper by Stone (1977) showed the important result that several nonparametric estimation methods are universally consistent.

Corder GW, Foreman DI. Nonparametric statistics for non-statisticians: a step-by-step approach. Hoboken (NJ): Wiley; 2009.

Györfi L, editor. Principles of nonparametric learning. New York: Springer; 2002.

Györfi L, Kohler M, Krzyzak A, Walk H. A distribution-free theory of nonparametric regression. New York: Springer; 2002.

Hardle W. Applied nonparametric regression. Cambridge: Cambridge University Press; 1992.

Nadaraya EA. Nonparametric estimation of probability densities and regression curves. Dordrecht: Kluwer; 1989.

Stone CJ. Consistent nonparametric regression. Ann Stat 1977;5:595–645.

CHAPTER 15

Learning Function Estimation

In Chapter 14, we described the formulation of the function estimation problem. In this chapter, we discuss methods for estimation. Specifically, we consider how the methods we have considered to this point for classification can be extended to estimation.

15.1 REVIEW OF THE FUNCTION ESTIMATION/REGRESSION PROBLEM

We begin by reviewing the formulation of the function estimation problem (vs classification) that we discussed earlier.

Instead of two classes (0 and 1), we wish to estimate a real value. That is, we want to find a function f such that $f(\overline{x})$ provides a good estimate of the value of y given a feature vector $\overline{x}$.

We assume that there is an unknown probability distribution (density) $p(\overline{x}, y)$ determining $p(\overline{x})$ and $p(y|\overline{x})$ for all relevant $\overline{x}$ and y. Since y can take on a continuum of values, trying to estimate y exactly is generally unreasonable and the criterion of minimizing the probability of error needs to be replaced by some other performance criterion.

We need to introduce a loss function (or a cost function) to measure how close our estimate $f(\overline{x})$ is to the observed value y. A best estimate is one that minimizes the expected cost (or *loss*). (There may be more than one best estimate.) Squared error loss $(y - f(\overline{x}))^2$ is the most common loss function and this is what we use.

Given any function f, for a given value of $\overline{x}$, the expected loss of $f(\overline{x})$ is

$$R(f, \overline{x}) = \int_{\mathbf{R}} (y - f(\overline{x}))^2 \, p(y|\overline{x}) \, dy.$$

An Elementary Introduction to Statistical Learning Theory, First Edition.
Sanjeev Kulkarni and Gilbert Harman.
© 2011 John Wiley & Sons, Inc. Published 2011 by John Wiley & Sons, Inc.

The expected loss of f averaged over all values of $\overline{x}$ is given by

$$R(f) = \int_{\mathbf{R}^d} R(f, \overline{x}) \, p(\overline{x}) \, d\overline{x} = \int_{\mathbf{R}^d} \int_{\mathbf{R}} (y - g(\overline{x}))^2 \, p(y|\overline{x}) \, p(\overline{x}) \, dy \, d\overline{x}.$$

As in classification, a Bayes rule is a mapping that minimizes the expected loss. For squared error loss, it can be shown that a best estimate (i.e., Bayes rule) is the conditional mean of y given $\overline{x}$. This is often called the *regression function*.

The Bayes loss (or Bayes error rate) R^* is the expected loss of a Bayes rule.

If we knew the underlying distributions we could compute a Bayes rule. However, as in the case of classification, we generally know little or nothing about these distributions. Instead, we assume that we have labeled data $(\overline{x}_1, y_1), (\overline{x}_2, y_2), \ldots, (\overline{x}_n, y_n)$.

Given this data, we wish to find a function with small error rate. Our benchmark (the best we can possibly do) is the Bayes rate R^*.

15.2 NEAREST NEIGHBOR RULES

Given a feature vector $\overline{x}$, the 1-NN nearest neighbor rule for classification finds the feature vector among the training examples $\overline{x}_i$ that is closest to $\overline{x}$ and assigns $\overline{x}$ the corresponding label. For estimation, the obvious extension is to let the corresponding label be our estimate for y. It turns out that for squared error loss, the 1-NN rule has asymptotic error rate *equal* to $2R^*$. This contrasts with the performance of the 1-NN rule for classification for which the asymptotic error rate is bounded by $2R^*$ but not necessarily equal to it.

The k-NN rule for classification considers the k nearest neighbors and takes the majority vote of the corresponding labels. A natural extension for k-NN estimation is to estimate the value of y at a given point $\overline{x}$ by taking the average of the k values of the y_i corresponding to the k examples $\overline{x}_i$ that are closest to $\overline{x}$.

As with classification, we can also consider an extension to the k_n-NN rule in which the number of neighbors k_n grows with the amount of training data n. An estimation result can be obtained under the same conditions as for classification, that is, if $k_n \to \infty$ and $k_n/n \to 0$ then the k_n-nearest neighbor rule is universally consistent.

15.3 KERNEL METHODS

For kernel methods, we can fix a distance h and estimate the value of y at a given point $\overline{x}$ by taking the average of all y_i such that $|\overline{x}_i - \overline{x}| \leq h$.

As with classification, this can be written in terms of the "window function" or kernel of Equation (8.1) described in Chapter 8:

$$K(\overline{z}) = \begin{cases} 1 & \text{if } |\overline{z}| \leq 1 \\ 0 & \text{otherwise.} \end{cases}$$

The number of points within distance h of $\overline{x}_i$ is given by

$$W = \sum_{i=1}^{n} K\left(\frac{\overline{x} - \overline{x}_i}{h}\right).$$

The estimate $f(\overline{x})$ is then given by

$$f(\overline{x}) = \frac{1}{W} \sum_{i=1}^{n} y_i K\left(\frac{\overline{x} - \overline{x}_i}{h}\right).$$

More generally, as before, we can work with other kernel functions. Many choices for $K(\cdot)$ are possible as long as certain fairly mild assumptions are satisfied. Some standard choices are triangular, Gaussian, and so on, as described in Chapter 8.

For general kernels, the expressions for W and $f(\overline{x})$ remain the same. Now W represents the total weights as opposed to the number of training examples within distance h of $\overline{x}$. The estimate $f(\overline{x})$ is then a weighted average of y_i, where the weights are determined by how close $\overline{x}_i$ is to $\overline{x}$, the kernel function $K(\cdot)$, and the smoothing factor h.

As with classification, to get a universally consistent estimator, we need to let $h = h_n$ be a function of the number of examples n. If $h_n \to 0$ and $nh_n^d \to \infty$, then the kernel estimate will be universally consistent.

15.4   NEURAL NETWORK LEARNING

In neural network learning for classification, we started with perceptrons in Chapter 9. A perceptron takes a weighted sum of the inputs and passes this weighted sum through a threshold. For multilayer networks discussed in Chapter 10, we replaced the hard threshold with a sigmoidal function (a sort of soft threshold). To get a binary classification, the final output was passed through a threshold unit.

The use of a sigmoidal activation rule lends itself naturally to estimation. The final output unit is *not* passed through a threshold, which allows the final output of the network to take on real values in an interval. Sometimes the activation rule for the final output unit is taken to be linear rather than sigmoidal. That is, the output of the final unit is just the weighted sum of the inputs. This allows representing functions that take on values from $-\infty$ to ∞ rather than over some finite interval.

Of course, the perceptrons other than the output unit still use a sigmoidal activation rule. The nonlinearity is necessary to approximate general functions. If all the perceptrons had a linear activation rule, then the entire multilayer network computes some linear rule and so the whole network could be reduced to a single unit. This would allow representing only linear rules, a limitation analogous to what we faced for single perceptrons in the case of classification.

However, as with classification, with the nonlinearity there is a universal approximation result for multilayer networks. In fact, only one hidden layer with enough units is required to obtain the approximation. Specifically, let $g(\overline{x})$ be any continuous function from $[-1, 1]^d$ into $\mathbf{R}$. Then for any $\epsilon > 0$ there exists an integer m and some constants $a_i \in \mathbf{R}$, $b_i \in \mathbf{R}$, $\overline{w}_i \in \mathbf{R}^d$ such that the function

$$f(\overline{x}) = \sum_{i=1}^{m} a_i \sigma(\overline{w}_i \cdot \overline{x} + b_i)$$

approximates g to within ϵ.

Here, $\overline{w}_i \cdot \overline{x}$ denotes the dot product (i.e., weighted sum) of the vector $\overline{w}_i$ with the vector $\overline{x}$, namely,

$$\overline{w}_i \cdot \overline{x} = w_{i,1} x_1 + \cdots + w_{i,d} x_d.$$

ϵ is a small number that represents how accurately we wish to approximate the continuous function g. Specifically, the resulting function f is such that $|f(\overline{x}) - g(\overline{x})| < \epsilon$ for all $\overline{x} \in [-1, 1]^d$. If we look at the form of f, we see that f is written as a weighted linear combination of m terms with weights a_i. The terms being combined are just the outputs of m units with the sigmoidal activation rule $\sigma(\cdot)$. It turns out that the nonlinearity and architecture of the network is what gives the universal approximation property. The specific shape of the sigmoid is not important nor even the fact that the activation rule is sigmoidal.

As with classification, being able to represent the general decision rules is only one of three issues we considered. The second issue is finding effective means for selecting the weights (i.e., training the network) so that a good rule can be selected from among the rules that can be represented by the network. As might be expected, backpropagation can be used for a learning algorithm for estimation.

The third issue we considered in classification using neural networks is how to get an idea of sample size bounds required for learning. We would like a PAC-type result for estimation similar to what we have for classification. We consider this next.

15.5 ESTIMATION WITH A FIXED CLASS OF FUNCTIONS

Let us suppose that a learning problem specifically involves a fixed class $\mathcal{F}$ of functions, with each $f \in \mathcal{F}$ being a map from the feature space, R^d, to R. $f(\overline{x})$ is the prediction for the y value when we observe feature vector $\overline{x}$. The learner's task is to select a good function from the $\mathcal{F}$.

The learner will see labeled examples $(\overline{x}_1, y_1), \ldots, (\overline{x}_n, y_n)$ and wishes to select a good function from the class $\mathcal{F}$, where the goodness of a function f is determined by its error rate $R(f)$, which we measure using the average squared error of f.

The minimum error rate over the class $\mathcal{F}$ is $R_{\mathcal{F}}^* = \min_{f \in \mathcal{F}} R(f)$. This is the best performance the learner can hope for, if the learner is restricted to rules from $\mathcal{F}$.

With a finite amount of training data, the learner cannot expect to find the best function from $\mathcal{F}$. The learner must therefore aim to select a function h from $\mathcal{F}$ such that the error rate $R(h)$ of the function satisfies $R(h) \leq R_{\mathcal{F}}^* + \epsilon$.

The learner cannot even expect *always* (i.e., for *any* set of observed examples) to produce an h that is approximately correct. But the learner can hope to probably do so.

Thus, similarly to the case of classification, we are led to the following PAC criterion for estimation:

$$\Pr\{R(h) \leq R_{\mathcal{F}}^* + \epsilon\} \geq 1 - \delta.$$

A class of functions $\mathcal{F}$ is PAC learnable if there is a mapping that produces a function $h \in \mathcal{F}$ based on training examples such that for every $\epsilon, \delta > 0$ there is a finite sample size $m(\epsilon, \delta)$ such that for any distributions we have

$$\Pr\{R(h) > R_{\mathcal{F}}^* + \epsilon\} < \delta$$

after seeing $m(\epsilon, \delta)$ training examples.

Some examples of function classes $\mathcal{F}$ of interest include the following:

- Linear estimates: Fitting a linear estimate to the data, for example, a straight line to a function of one feature.
- Higher order polynomials.
- Neural networks with a fixed architecture.

A natural strategy to select a function from $\mathcal{F}$ is to choose one that fits the data the best (i.e., has minimal empirical loss on the data). This is called *empirical risk minimization*. This is a generalization of what is sometimes called *enumerative induction* in the case of classification.

As before, empirical risk minimization works (and in fact the class $\mathcal{F}$ is PAC learnable) if the set $\mathcal{F}$ is not too "rich." In the case of classification, richness of the class of decision rules $\mathcal{C}$ was measured by its VC-dimension. In the next section, we describe an extension for the case of classification.

15.6 SHATTERING, PSEUDO-DIMENSION, AND LEARNING

Consider a set of points $(\overline{x}_1, y_1), \ldots, (\overline{x}_n, y_n)$. A function f that does not pass through any of these points splits the points into two subsets: those that lie below the function (i.e., $y_i < f(\overline{x}_i)$) and those that lie above the function (i.e., $y_i > f(\overline{x}_i)$). We will say that these points get labeled 0 and 1, respectively.

The set of points $(\overline{x}_1, y_1), \ldots, (\overline{x}_n, y_n)$ is *shattered* by a class of functions $\mathcal{F}$ if all 2^n labelings of these points can be generated using functions from $\mathcal{F}$.

The *pseudo-dimension* of a class of functions $\mathcal{F}$, denoted $\dim(\mathcal{F})$, is the largest integer V such that *some* set of V points is shattered by $\mathcal{F}$. If arbitrarily large sets can be shattered, then $\dim(\mathcal{F}) = \infty$. Pseudo-dimension is a generalization of VC-dimension to real-valued functions.

Under certain mild regularity conditions, if $\dim(\mathcal{F}) < \infty$ then $\mathcal{F}$ is PAC learnable. As with classification, sample size bounds that depend on $\dim(\mathcal{F})$, ϵ, and δ are also available.

However, for estimation the converse is *not* true. That is, there are examples where $\mathcal{F}$ has infinite pseudo-dimension, but $\mathcal{F}$ is still PAC learnable. This contrasts with the case of classification; for which (under mild regularity assumption), $\mathcal{C}$ is PAC learnable if and only if $\mathcal{C}$ has finite VC-dimension.

There are some refined results, which include converse results, involving a notion of "fat shattering." Rather than requiring the functions f to simply go above or below the points $(\overline{x}_i, y_i)$ to shatter them, we require f to go above or below by at least some fixed amount $\gamma > 0$ (Figure 15.1).

Specifically, given a function f, the point $(\overline{x}_i, y_i)$ will be labeled 0 if

$$y_i < f(\overline{x}_i) - \gamma$$

and $(\overline{x}_i, y_i)$ will be labeled 1 if

$$y_i > f(\overline{x}_i) + \gamma.$$

The set of points $(\overline{x}_1, y_1), \ldots, (\overline{x}_n, y_n)$ is said to be γ-*shattered* by a class of functions $\mathcal{F}$ if all 2^n labelings of these points can be generated using functions from $\mathcal{F}$. $\dim_\gamma(\mathcal{F})$ denotes the largest integer V such that some set of V points is

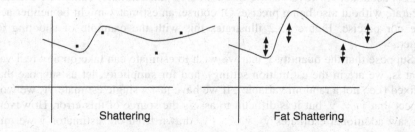

Figure 15.1 Shattering and fat shattering: (a) shattering and (b) fat shattering.

γ-shattered by $\mathcal{F}$. Some refined and converse results can be obtained in terms of finiteness of $\dim_{\gamma}(\mathcal{F})$.

15.7 CONCLUSION

In this chapter, we discussed extensions of classification methods and results described in previous chapters for the estimation problem. Nearest neighbor and kernel methods extended in a natural way by suitably averaging the labels y_i instead of voting as was done in the case of classification. Universal consistency results hold for estimation under similar conditions as for classification. Multilayer neural networks can be used for estimation in the natural way by using sigmoidal activation rules, although the final output unit can use a simple linear activation rule. Arbitrary continuous functions can be approximated by multilayer networks, and backpropagation can be used to train such networks. The results of PAC learning can be extended to the estimation problem by considering a suitable notion of shattering and pseudo-dimension, which generalize the notion of shattering and VC-dimension. For estimation, if a class of functions $\mathcal{F}$ has finite pseudo-dimension then it is PAC learnable, but the converse is not true in general, though some converse results can be obtained using a notion of fat shattering and associated dimensions.

15.8 APPENDIX: ACCURACY, PRECISION, BIAS, AND VARIANCE IN ESTIMATION

Colloquially, accuracy and precision are often used with similar meanings, but in statistics (and math, science, and engineering more generally) the two terms have very distinct meanings. Accuracy refers to the degree to which a measurement or estimate is close to an actual value or truth. Precision refers to the reproducibility of a measurement or estimate.

A common example for illustrating the difference between accuracy and precision uses the example of shooting at a target. An estimator can be very precise but not accurate. However, as we discuss below, an estimator cannot be extremely accurate without also being precise. Of course, an estimator might be neither accurate nor precise. Figure 15.2 illustrates this with the example of shooting at a target.

Suppose that the quantity y that we wish to estimate can take on any real value (that is, we are in the estimation setting), and for simplicity, let us suppose that y is fixed (i.e., not a random variable). If we have just a single estimate $\hat{y}_1$, we would expect that $\hat{y}_1 \neq y$, but it is difficult to assess the source of this error. However, if we saw additional estimates $\hat{y}_2, \hat{y}_3, \ldots, \hat{y}_k$ drawn from our estimator $\hat{y}$ we might get some idea of the types of errors produced by the estimator. One source of

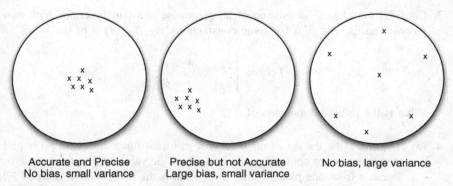

| Accurate and Precise | Precise but not Accurate | No bias, large variance |
| No bias, small variance | Large bias, small variance | |

Figure 15.2 Accuracy = bias + variance: (a) accurate and precise—no bias, small variance; (b) precise but not accurate—large bias, small variance; and (c) no bias, large variance.

error is the variability in the estimates $\hat{y}_1, \hat{y}_2, \ldots, \hat{y}_k$. This can be measured by the variance of the estimator $\hat{y}$ about its mean. The variance of an estimator is one way to measure precision.

The second source of error is called the bias of the estimator $\hat{y}$, which is a measure of how far away the mean of the estimator is from the underlying quantity y that we wish to estimate. The accuracy of an estimator is related to its bias and variance. One measure of the accuracy of an estimator is simply its average (or expected) squared error, which is what we have been using. That is to say, $E[(y - \hat{y})^2]$. But since y is assumed to be fixed, this can be written as

$$E[(y - \hat{y})^2] = [y - E(\hat{y})]^2 + \sigma^2(\hat{y}).$$

The first term is the bias of the estimator and the second term is its variance. Measured in this way, the accuracy of the estimator is the sum of its bias and variance. Note that both terms are positive, so they cannot cancel each other out. Hence, an estimator that lacks precision (i.e., has a large variance), cannot be very accurate even if its bias is zero.

15.9 QUESTIONS

1. In function estimation (as opposed to classification), finiteness of which of the following is sufficient for PAC learning? VC-dimension; pseudo-dimension; Popper dimension; dimension-X; fractal dimension; feature space dimension; Hausdorff dimension.

2. What is the pseudo-dimension of all constant functions? That is, $f \in \mathcal{F}$ if $f(x)$ is of the form

$$f(x) = \alpha \quad \text{for some constant } \alpha.$$

3. Consider the class $\mathcal{F}$ consisting of all piecewise constant functions with two pieces. That is, $f \in \mathcal{F}$ if for some constants $\alpha_1, \alpha_2, \beta, f(x)$ is of the form

$$f(x) = \begin{cases} \alpha_1 & x \le \beta \\ \alpha_2 & x > \beta. \end{cases}$$

What is the pseudo-dimension of $\mathcal{F}$?

4. (a) If we let $\mathcal{F}$ be the set of *all* piecewise constant functions (with no bound on the number of constant pieces and where the value of $\mathcal{F}$ may increase or decrease from one piece to the next), what is the pseudo-dimension of $\mathcal{F}$?

   (b) Do you think this class of functions is PAC learnable? Why or why not? (You do *not* need to give a proof. Just briefly provide some intuitive justification for your answer.)

   (c) Does your answer in the previous part follow from the learnability result for estimation? Why or why not?

5. Consider the class $\mathcal{F}$ of all nondecreasing functions of a single variable x. That is, a function $f(x)$ belongs to $\mathcal{F}$ if $f(x_2) \ge f(x_1)$ whenever $x_2 > x_1$. What is the pseudo-dimension of $\mathcal{F}$? Justify your answer.

15.10 REFERENCES

Cover (1968) showed an early result that the asymptotic error rate of the 1-NN rule for estimation is $2R^*$. Universal consistency of nearest neighbor, kernel, and other rules was shown by Stone (1977) and has been further studied and refined by many researchers since then. The book by Györfi *et al.* (2002) contains an advanced treatment of many results in the area. The fact that multilayer networks with sigmoidal activation rules are universal approximators of functions was shown by Cybenko (1989). The ideas of pseudo-dimension, fat shattering, and other associated dimensions and learning results have been studied and used by Pollard (1984), Haussler (1992), Alon *et al.* (1993), Kearns and Schapire (1994), Bartlett *et al.* (1996), and others.

Alon N, Ben-David S, Cesa-Bianchi N, Haussler D. Scale sensitive dimensions, uniform convergence, and learnability. Symposium on Foundations of Computer Science. Washington, D.C.: IEEE Computer Society Press; 1993.

Bartlett PL, Long PM, Williamson RC. Fat-shattering and the learnability of real-valued functions. J Comput Syst Sci 1996;52(3):434–452.

Cover TM. Estimation by the nearest neighbor rule. IEEE Trans Inf Theory 1968;IT-14:50–55.

Cybenko G. Approximations by superpositions of sigmoidal functions. Math Control Signals Syst 1989;2(4):303–314.

Györfi L, Kohler M, Krzyzak A, Walk H. A distribution-free theory of nonparametric regression. New York: Springer; 2002.

Haussler D. Decision theoretic generalizations of the PAC model for neural net and other learning applications. Inf Comput 1992;100:78–150.

Kearns MJ, Schapire RE. Efficient distribution-free learning of probabilistic concepts. J Comput Syst Sci 1994;48(3):464–497.

Pollard D. Convergence of stochastic processes. New York: Springer; 1984.

Stone CJ. Consistent nonparametric regression. Ann Stat 1977;5:595–645.

CHAPTER 16

Simplicity

16.1 SIMPLICITY IN SCIENCE

Scientists and other investigators often say that one hypothesis is preferable to another because it is simpler. It is interesting to consider how such appeals relate to the misfit versus complexity trade-off we discussed in Chapter 13.

There are at least two ways in which simplicity seems relevant to scientific reasoning—explicitly and implicitly. Sometimes scientists make explicit appeals to simplicity. On the other hand, simplicity can play an implicit role in theory choice by limiting the hypotheses that scientists take seriously, without anyone needing to explicitly rule out certain hypotheses as too complicated. This second role is different from the first, because when scientists explicitly appeal to simplicity in favor of one theory over another, both theories are being taken seriously.

16.1.1 Explicit Appeals to Simplicity

We are concerned with the second more implicit role of simplicity, but it will help to say a few words about the first case, in which an explicit appeal is made to one or another type of simplicity, for instance, when a scientist argues that theory T_1 is to be preferred to theory T_2 because the principles of T_1 are simpler in certain ways. Or it might be said that T_1 has a simpler "ontology" than T_2. For example, T_1 and T_2 might be theories in physics and T_1 appeals to a smaller number of basic subatomic particles. In this case, there may be mention of the principle of "Occam's Razor," which says we should not multiply (kinds of) entities beyond necessity.

Or, someone might argue that T_1 is simpler than T_2 in the sense that it postulates fewer coincidences or explains the data with fewer *ad hoc* assumptions than T_2, so that the explanations provided by T_1 are simpler than those provided by T_2.

An Elementary Introduction to Statistical Learning Theory, First Edition.
Sanjeev Kulkarni and Gilbert Harman.
© 2011 John Wiley & Sons, Inc. Published 2011 by John Wiley & Sons, Inc.

16.1.2 Is the World Simple?

It might be objected that any such appeal to simplicity wrongly assumes that the world is simple. Surely, we have no reason to think that the world is simple, and in fact have good reason to think that the world is quite complicated.

However, such an objection is mistaken. To appeal to simplicity is not to say we have to choose a simple hypothesis. What is being appealed to is *relative* simplicity, not *absolute* simplicity. Both hypotheses can be very complicated even though one of them is simpler than the other.

As we have seen in Chapter 13, trading off considerations of fit with considerations of something like simplicity can support a kind of learning in the limit. That way of taking "simplicity" into account does not suppose that "the world is simple"—it works whether or not the world is simple.

16.1.3 Mistaken Appeals to Simplicity

A different sort of objection goes back to the point already noted that, when a scientist explicitly appeals to simplicity in favor of one hypothesis over another, there are reasons to take both hypotheses seriously. In such a case, there may or may not be a good reason to think that the simpler hypothesis is more likely. Sometimes, the simpler hypothesis is not more likely.

Consider a situation in evolutionary biology in which two kinds of animal share a common feature. Hypothesis 1 says that they share this feature because of inheritance from a common ancestor. Hypothesis 2 says that they did not inherit the feature from a common ancestor but instead there was a parallel development in their ancestors. The fact that Hypothesis 1 is in a certain sense simpler than Hypothesis 2 is not necessarily a good reason to prefer Hypothesis 1 to Hypothesis 2. It is relevant how likely such parallel development is, given the benefits of having that feature. The "wings" of birds and bats are not due to the wings of a common ancestor but are the result of such parallel development. Under certain conditions, Hypothesis 2 is more likely than Hypothesis 1 despite the apparent simplicity of Hypothesis 1 (Sober, 1990).

16.1.4 Implicit Appeals to Simplicity

It is easy to see that scientists must in some sense at least implicitly appeal to considerations of simplicity. In any interesting scientific cases, the data will now and forever be in accord with infinitely many competing hypotheses. If scientists are ever to reach even tentative conclusions, they must somehow discriminate among those hypotheses.

In other words, infinitely many hypotheses that fit the data as well as a preferred hypothesis h must be set aside at any given stage. This is not to say that one of these discarded hypotheses, say u, must forever be set aside. u may be brought back into consideration later if h fails to fit new data in a way that one of the previously set aside hypotheses does.

For example, when scientists consider how bodies attract each other, they might be led to the hypothesis h that the force of attraction is directly proportional to the product of the masses of the bodies and inversely proportional to the square of the distance r between them, $F = Gm_1m_2/r^2$. Now the evidence so far does not distinguish between this hypothesis and the hypothesis u that this principle is correct up until today but that tomorrow the correct formula will depend on the cube of the distance, as in $F = Gm_1m_2/r^3$. But no scientist would even consider this last possibility and would not take seriously anyone who put it forward as an alternative to the first hypothesis. That is, no one would take this seriously given our present evidence. If tomorrow, we started getting evidence conflicting with the first square hypothesis and in accord with the second cube hypothesis, then we might reluctantly reconsider the exclusion of the cube hypothesis.

16.2   ORDERING HYPOTHESES

It seems then that ordinary scientific practice must somehow impose an ordering on hypotheses that places h ahead of u. Looking at the actual ordering that scientists use, it seems that this ordering has something to do with the intuitive simplicity of a hypothesis. The hypothesis h is intuitively simpler than the hypothesis u.

Similarly, scientists consider linear equations before they consider quadratic equations, which in turn are taken seriously before cubic equations are. Putnam (1963) claims that standard scientific practice looks something like this. Given a collection of data D_1, scientists think of a few hypotheses that might explain D_1, namely, $h_1, h_2, \ldots, h_k$. (Many more hypotheses might explain D_1 which are not thought of or, if thought of, are not taken seriously.) Scientists next look for data D_2 that would eliminate all but one of the h_i. They then think up a few additional hypotheses that might explain D_1 and D_2. They repeat this indefinitely.

How is simplicity involved in this practice? Perhaps scientists tend to think of simpler hypotheses first. Or perhaps to be a simple hypothesis is just to be the sort of hypothesis scientists tend to think of first (Nozick, 1983).

16.2.1   Two Kinds of Simplicity Orderings

It is useful to distinguish at least two ways in which hypotheses might be ordered in terms of simplicity: There might be a well ordering of individual hypotheses

$$h_1, h_2, h_3, \ldots, h_k, \ldots$$

in the sense that (a) for any two different hypotheses h_i and h_j either h_i is ordered before h_j or h_j is ordered before h_i and (b) for any nonempty subset of hypotheses there is always a simplest one (i.e., first in the ordering). That is one way in which hypotheses might be ordered in terms of simplicity. Another possibility is that there is an ordering of inclusive classes of hypotheses

$$H_1 \subset H_2 \subset H_3 \subset \cdots \subset H_k \subset \cdots$$

that does not suppose that individual hypotheses are well ordered. The misfit versus complexity trade-off discussed in Chapter 13 was of this second type.

The first type of simplicity ordering, involving a well ordering of hypotheses themselves, might order the hypotheses in terms of the length of certain representations, with hypotheses of the same length ordered "alphabetically." In some approaches, the hypotheses are represented as computer programs of a certain sort, and are ordered according to the length of the computer program. This type of conception of simplicity based on minimum program length is often appealed to in computer science and often goes by the term *Kolmogorov complexity*. It is related to a certain notion of "randomness," according to which a sequence of numbers is random to the extent that there is no significantly shorter way of expressing the sequence than by simply listing the numbers in the sequence.

The second type of simplicity ordering orders classes of hypotheses rather than individual hypotheses. Hypotheses are sometimes grouped in terms of the parameters needed to specify particular members of the group. Linear hypotheses are ordered before quadratic hypotheses, for example, because in the two-dimensional case, two parameters are sufficient to determine a linear hypothesis of the form $ax + b$, whereas three are needed to determine a quadratic hypothesis of the form $ax^2 + bx + c$.

The first approach allows only countably many hypotheses to be considered, since the hypotheses themselves are well ordered. The second approach can allow for uncountably many hypotheses since each of the classes in the ordering may itself contain uncountably many hypotheses. (The meaning of "uncountably many" is discussed in Section 4.7.) However, the second approach can be used even if there are only countably many hypotheses being considered (such as those represented in some fixed alphabet). In this case, even though there are only countably many hypotheses, they need not be well ordered. For example, consider the set of polynomials with rational coefficients. There are only countably many of these, but using the second approach, we may wish to order all linear hypotheses (infinitely many) before nonlinear quadratic hypotheses, and these before third-order polynomials, and so on.

16.3 TWO EXAMPLES

16.3.1 Curve Fitting

Sometimes investigators want to fit a curve to some data points. In a two-dimensional example, we might let the y coordinate represent the value of a quantity we assume to be a function of a quantity represented by the x coordinate. Suppose our data indicate values of y for certain values of x, as in Figure 16.1.

Various curves go through these four points. One is given by the formula $y = 2x$, as in Figure 16.2. Another is given by the formula

$$y = 2x + (x - 1)(x - 3)(x - 4)(x - 6),$$

as in Figure 16.3.

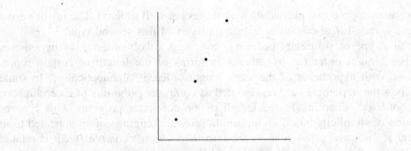

Figure 16.1   Data points (1, 2), (3, 6), (4, 8), (6, 12).

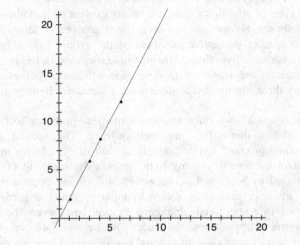

Figure 16.2   Curve for $y = 2x$.

Which curve provides a better hypothesis about the relation between x and y?

In a standard curve-fitting context, investigators would opt for the linear relationship $y = 2x$ over the considerably more complex relationship

$$y = 2x + (x - 1)(x - 3)(x - 4)(x - 6).$$

16.3.2   Enumerative Induction

Suppose the data consist in results of examining emeralds for color. The data have been obtained before a certain time T and all emeralds examined have been found to be green. Consider the following two hypotheses:

h_1 All emeralds are green.

h_2 All emeralds are either first examined for color before T and green or not so examined and blue.

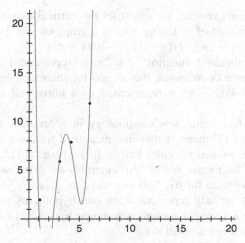

Figure 16.3 Curve for $y = 2x + (x - 1)(x - 3)(x - 4)(x - 6)$.

Both hypotheses fit the data equally well. Furthermore, the hypotheses make conflicting claims about emeralds that are not first examined before T.

However, it is clear that in this circumstance an investigator would take only h_1 seriously and would not consider h_2 except in a philosophy class!

How are we to characterize the distinction here between hypotheses that are to be taken seriously and hypotheses that are not to be taken seriously?

A natural suggestion is that we are not to take seriously a hypothesis if there is a much *simpler* hypothesis that accounts for the data equally well.

Two questions immediately arise. First, how can we characterize the simplicity involved? Second, why should there be a preference for simpler hypotheses? Let us start by considering the first issue, how to characterize the relevant sort of simplicity.

16.4 SIMPLICITY AS SIMPLICITY OF REPRESENTATION

One natural suggestion is that the simplicity of a hypothesis has to do with the simplicity of our representation of that hypothesis. This accounts for the cases we have been considering.

In our previous example, where we compared two curves, $y = 2x$ and $y = 2x + (x - 1)(x - 3)(x - 4)(x - 6)$, the formula for the second curve is more complex than the formula for the first curve. The first curve is a straight line (Figure 16.2) whereas the second curve has a couple of ups and downs (Figure 16.3).

Similarly, the statement above of hypothesis h_1 about the color of emeralds is considerably shorter and simpler than the statement of hypothesis h_2.

There are, however, other ways to represent these hypotheses than the ones so far considered. How a hypothesis is represented on a graph depends on the coordinates used. For example, we might let the vertical axis represent the log of y or some other function of y. Figure 16.4 is a graph of $y = 2x$ when the vertical axis represents $y/(2x + (x - 1)(x - 3)(x - 4)(x - 6))$.

Here, the hypothesized function $y = 2x$ is represented by a complicated curve. Using the same coordinates, the second function we considered, $y = 2x + (x - 1)(x - 3)(x - 4)(x - 6)$, is represented by a horizontal straight line with a constant value of 1.

We might introduce some new terminology in order to restate hypotheses h_1 and h_2. Let "grue (at t)" mean "either first examined for color before T and green (at t), or not first examined for color before T and blue (at t)." Then the complex hypothesis $H2$ can be formulated as "All emeralds are grue" which is as simple a representation as we used for h_1, "All emeralds are green."

In fact, we have already represented the two hypotheses very simply as "h_1" and "h_2." If that is not simple enough, any single hypothesis could be represented by a single letter or even a small dot.

Mere simplicity of representation is not enough.

16.4.1 Fix on a Particular System of Representation?

One way to avoid such problems would be to fix the system of representation ahead of time and then take seriously only hypotheses that can be expressed in relatively few symbols.

This would rule out changing the graphs of the functions $y = 2x$ and $y = 2 + (x - 1)(x - 3)(x - 4)(x - 6)$. The first function would always be represented more simply than the second and would always count as simpler than the second, which seems right. Similarly, we could not use new terms like "grue" and "bleen" to

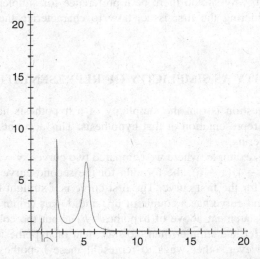

Figure 16.4 Vertical axis represents $y/(2x + (x - 1)(x - 3)(x - 4)(x - 6))$.

simplify hypotheses, which would block the bad result about the color of emeralds. And we could not allow hypotheses to be represented as "h_1" or "h_2."

But this way of handling simplicity is overly conservative. It overlooks the fact that sometimes we need to use new symbols to express new sorts of functions. For example, at some point people wanted to introduce symbols for trigonometric functions, or symbols for derivatives and integrals in calculus in order to express in a simple way hypotheses that they wanted to take seriously. If they had to represent those hypotheses without such new symbols, the hypotheses would have had very complex, perhaps even infinite, representations and would count as very complex, if simplicity is measured by simplicity of representation.

16.4.2 Are Fewer Parameters Simpler?

We mentioned above that it would be a mistake to fix a particular system of representation for all time and measure the simplicity of a hypothesis by the simplicity of its representation in that system. Progress in mathematics and science sometimes requires new notations, as in trigonometry and calculus, that change the simplicity with which we can represent certain hypotheses.

Trigonometric functions also provide an example of an objection to a simplicity measure in terms of the number of parameters needed to specify a function from a given class. Consider the class of functions, $a \sin(bx) + c$. Three parameters are enough to specify a particular function from this class. But, however many data points there are, it will almost always be possible to find some sine curve of this form that goes through every one of the points. The problem is that the sine curve can have a very high frequency. But, in general, it is not rational always to prefer a sine function from that class over a polynomial in a class of functions requiring four parameters to be specified.

Observe that the pseudo-dimension of the class of sine curves is infinite, even though only three parameters are needed to specify a particular member of this class. From the vantage point of Chapter 15, we can see that the number of parameters needed to pick out a member of a set is not always a good measure of the complexity of the set.

16.5 PRAGMATIC THEORY OF SIMPLICITY

Another consideration is how simple or easy it is to use a theory to answer questions in which we are interested. Suppose our question concerns the color of a particular emerald. It is very easy to use h_1 to get an answer to our question, since h_1 tells us that the answer is always going to be "green" no matter which emerald is in question. On the other hand, it can be less easy to use h_2 to answer this question, because we need to find out when the emerald is first observed. If the emerald is first observed before T, we can use h_2 to answer that the emerald is green; if the emerald is not first observed before T, we can use h_2 to answer that the emerald is blue.

For another example, suppose we have a function $y = f(x)$ and we want to know what is the value of this function when $x = 5$. If the function is $y = 2x$, then our calculation is relatively easy $y = 2(5) = 10$. On the other hand, if the function is

$$y = 2x + (x - 1)(x - 3)(x - 4)(x - 6)$$

then the calculation is more complex:

$$y = 2(5) + (5 - 1)(5 - 3)(5 - 4)(5 - 6) = 10 + (4)(2)(1)(-1) = 10 - 8 = 2.$$

Notice that changing to a new system of graphing coordinates or adding terms such as "grue" and "bleen" will not normally make such calculations easier. In fact, our calculation will often involve first substituting for the newly defined terms to get to the original representation of the problem and then solving, which is more complicated. This suggests that simpler theories may be theories that are easier to use to get answers to questions in which we are interested.

16.6 SIMPLICITY AND GLOBAL INDETERMINACY

Global indeterminacy arises in comparing the hypothesis that you live in a world of tables and chairs with the hypothesis that you are a brain in a vat being given experiences as if you are in a world of tables and chairs. You would have exactly the same experience in either case, even in the long run, so nothing in your experience could distinguish these hypotheses.

There is the hypothesis that the universe has existed at least since the "Big Bang" billions of years ago, in contrast to the hypothesis that the universe was created by God a few thousand years ago, complete with fossils and evidence of an earlier existence. Again, these hypotheses might be equally compatible with all the experience you are ever going to get.

One might argue that in such cases we are justified in choosing the simplest of the competing hypotheses, given their empirical equivalence, where simplicity is measured by ease of use. In these cases, our normal hypothesis is compared with another hypothesis that is parasitic on the normal hypothesis, meaning that it will be somewhat easier to use the normal hypothesis. To use the parasitic hypothesis to make a prediction, you need to reason like this:

"This hypothesis says that for such and such reasons it is as if the other hypothesis were true. The other hypothesis predicts so and so. Therefore, it is as if so and so."

The reasoning or calculation needed to use a parasitic hypothesis will include the reasoning or calculation needed to use the normal hypothesis with an additional extra wrapper around that calculation. So the parasitic calculation has to be more extensive than the normal calculation, which means it will be less simple, which justifies our not taking such parasitic hypotheses seriously.

16.7   SUMMARY

We noted that simplicity is important in determining what hypotheses scientists take seriously in order to account for data. We have briefly considered how to characterize the relevant sort of simplicity. After considering and rejecting simplicity of representation as the relevant basis of hypothesis simplicity, we speculated that the relevant sort of simplicity may have something to do with the ease with which a hypothesis might be used for scientific purposes. We noted the difference between ordering individual hypotheses and ordering classes of hypotheses. Although it is sometimes thought that the relevant classes can be ordered in terms of the number of parameters needed to specify a particular member of the class, we noted that the class of sine curves is a counterexample to that thought.

16.8   APPENDIX: BASIC SCIENCE AND STATISTICAL LEARNING THEORY

In this chapter, we have noted that scientists seek simplicity in their theories and we have compared the scientific appeal to simplicity with the misfit versus complexity trade-off that can play a role in statistical learning. However, we do not mean to suggest that the scientific appeal to simplicity is the same thing as the misfit versus complexity trade-off in statistical learning theory or that scientists can use the techniques of statistical learning theory to come up with basic scientific theories.

For one thing, basic scientific theories (such as Newtonian mechanics, Einstein's special and general relativity, and quantum field theory) are not directly ways of classifying items on the basis of their observable features. Far from it. Such scientific theories tend to be related to observational consequences only by way of "controlled experiments," given various "auxiliary assumptions" about measuring devices, isolation from interference of various other factors, and so on. Statistical learning theory is concerned with finding certain patterns and regularities in the observable data. Even if the methods of statistical learning theory find certain regularities in connection with a given type of controlled experiment, scientists may want to uncover the "hidden laws" behind these regularities.

There are of course questions of limited resources. Should we invest in basic science or in engineering and applied science that does not require finding the hidden underlying laws? Should we invest in medical research or basic biology? Do we want to predict eclipses or understand and learn what really exists in the heavens?

One might defend the funding of basic science at least in part because basic science is intrinsically interesting quite apart from possible applications. One might also argue that basic science sometimes turns out to have important technologically useful applications that could not have been predicted ahead of time.

Technology has sometimes advanced without advances in basic science. For example, gunpowder, paper, and block printing were used in China long before they were known in Europe. Indian astronomy was quite accurate in predicting heavenly events like eclipses. But some technology has required basic understanding. (The methods of statistical learning theory would not by themselves have anticipated nuclear fission and fusion.)

16.9 QUESTIONS

1. Suppose that competing hypotheses h_1 and h_2 fit the data equally well but h_1 is simpler than h_2. Does a policy of accepting h_1 rather than h_2 in such a case have to assume that the world is simple? Why or why not?

2. Critically assess the following argument. "The hypothesis that the world is just a dream is simpler than the hypothesis that there really are physical objects. The two hypotheses account equally well for the data. So, it is more reasonable to believe that the world is just a dream than that there really are physical objects."

3. How could there be nondenumerably many hypotheses? Could a nondenumerable set of things be well ordered?

4. Does the hypothesis that all emeralds are grue imply that emeralds will change color in 2050?

5. Is it reasonable to use simplicity to decide among hypotheses that could not be distinguished by any possible evidence?

6. How should scientists measure the simplicity of hypotheses? Should a scientist favor simpler hypotheses over more complex hypotheses that fit the data equally well? Are there alternatives?

7. Is it reasonable for you to believe that you are not a brain in a vat being given the experiences of an external world? Explain your answer.

8. What is the *instrumentalist conception of theories*?

9. It is sometimes suggested that in selecting a hypothesis on the basis of data, one should balance the empirical error of hypotheses against their simplicity. Does statistical learning theory provide a critique of that suggestion?

16.10 REFERENCES

Ludlow (1998) offers a pragmatic account of simplicity, citing earlier versions in Peirce (1931–1958) and Mach (1960). Harman (1999) argues for a similar view after surveying alternatives. Related computational approaches to simplicity are defended in Angluin and Smith (1983), Blum and Blum (1975), Blum (1967), Gold (1967), Kugel (1977), Solomonoff (1964), Turney (1988), and Valiant (1979). Sober (1975) argues for a "semantic" interpretation of simplicity, which measures the complexity of a class of hypotheses in terms of the number of parameters needed to specify a particular member of the class. Later Sober (1988, 1990) argues against the general relevance to inference of certain notions of simplicity. Still later, Forster and Sober (1994) argue that simplicity (as measured by the paucity of adjustable parameters) helps maximize the goal of predictive accuracy.

The grue–bleen example comes from Goodman 1965).

Angluin DC, Smith CH. Inductive inference: theory and methods. Comput Surv 1983;15:237–269.

Blum M. A machine-independent theory of the complexity of recursive functions. J Assoc Comput Mach 1967;14:322–336.

Blum L, Blum M. Toward a mathematical theory of inductive inference. Inf Control 1975;28:125–155.

Dewey J. Logic: the theory of inquiry. New York: Holt; 1938.

Forster MR, Sober E. How to tell when simpler, more unified, or less ad hoc theories will provide more accurate predictions. Br J Philos Sci 1994;45:1–35.

Gigerenzer G, et al. Simple heuristics that make us smart. Oxford: Oxford University Press; 1999.

Gold EM. Language identification in the limit. Inf Control 1967;10:447–474.

Goodman N. Fact, fiction, forecast. 2nd ed. Indianapolis (IN): Bobbs-Merrill; 1965.

Harman G. Simplicity as a pragmatic criterion for deciding what hypotheses to take seriously. In: Harman G, editor. Reasoning, meaning, mind. Oxford: Oxford University Press; 1999.

Kugel P. Induction, pure and simple. Inf Control 1977;35:276–336.

Ludlow P. Simplicity and generative grammar. In: Stainton R, Murasugi K, editors. Philosophy and linguistics. Boulder (CO): Westview Press; 1998.

Mach E. The science of mechanics. 6th ed. Chicago: Open Court; 1960.

Mitchell TM. Machine learning. New York: McGraw-Hill; 1997.

Nozick R. Simplicity as fall-out. In: Cauman L, Levi I, Parsons C, editors. How many questions: essays in honor of Sydney Morgenbesser. Indianapolis: Hackett Publishing; 1983. pp. 105–119.

Peirce CS. Hartshorne C, Weiss P, Burks A, editors. Collected papers of Charles Sanders Peirce, 8 vols. Cambridge: Harvard University Press; 1931–58.

Putnam H. "Degree of confirmation" and inductive logic. In: Schillp A, editor. The philosophy of Rudolph Carnap. LaSalle, (IL): Open Court; 1963.

Sober E. Simplicity. Oxford: Oxford University Press; 1975.

Sober E. Reconstructing the past. Cambridge (MA): MIT Press; 1988.

Sober E. Let's razor Occam's razor. In: Knowles D, editor. Explanation and its limits. Cambridge: Cambridge University Press; 1990.

Solomonoff RJ. A formal theory of inductive inference. Inf Control 1964;7:1–22, 224–254.

Stalker D. Grue: the new riddle of induction. Peru (IL): Open Court; 1994.

Turney PD. Inductive inference and stability [PhD Dissertation]. Department of Philosophy, University of Toronto; 1988.

Valiant LG. The complexity of enumeration and reliability problems. SIAM J Comput 1979;8:410–421.

CHAPTER 17

Support Vector Machines

Support vector machines (SVMs) provide a state-of-the-art learning method that has been highly successful in a variety of applications. SVMs are a special case of the kernel methods discussed in Chapter 8. The success of SVMs has resulted in a great deal of attention to this particular form of kernel methods.

The origin of SVMs dates back to the work of Vapnik in the late 1970s, and arises from two key ideas. The first idea is to map the feature vectors in a nonlinear way to a high (possibly infinite) dimensional space and then utilize linear classifiers in this new space. This results in nonlinear classifiers in the original space, which overcomes the representational limitations of linear classifiers. However, the use of linear classifiers (in the transformed space) lends itself to computational methods for finding a classifier that performs well on the training data. These computational benefits are retained while allowing a rich class of nonlinear rules. The second key idea is that of a large-margin linear classifier that separates the data as much as possible from among the generally infinitely many hyperplanes that may separate the data. While many separating hyperplanes perform equally well on the training data (in fact perfectly well if they separate the data), the generalization performance on new data can vary significantly. The idea of choosing a separating hyperplane with a "large margin" helps provide good generalization performance.

We came across similar considerations in our discussion of neural networks in Chapters 9 and 10. Recall that in Chapter 9, we started with linear classifiers (namely, perceptrons), and we described a training rule to adjust the weights of the perceptron based on a set of training examples. This training rule is guaranteed to find a set of weights that correctly classifies all of the training examples whenever it is possible to do so. The problem is that perceptrons are able to represent only linear rules.

This motivated the consideration of multilayer networks in Chapter 10. We argued that, with three layers and enough perceptrons per layer, these networks can approximate general decision rules as closely as we like. The backpropagation

An Elementary Introduction to Statistical Learning Theory, First Edition.
Sanjeev Kulkarni and Gilbert Harman.
© 2011 John Wiley & Sons, Inc. Published 2011 by John Wiley & Sons, Inc.

training algorithm makes this a practical approach to dealing with the limitations of perceptrons (i.e., linear rules).

So, the problem of limited representation was addressed by considering multilayer networks, and the problem of a training algorithm was addressed by backpropagation. A third problem was that of ensuring that the resulting classifier has good performance. With multilayer networks, this is done by making sure that the size of the network (the number of layers and the number of perceptrons per layer) is not too large compared with the number of training examples. The bounds based on VC-dimension discussed in Chapter 12 give some guidance in this regard.

In this chapter, these problems are addressed in different ways. SVMs use linear rules, but only after mapping to a high-dimensional space. This results in nonlinear rules in the original space, which addresses the representational issue. The computational issue is addressed by formulating the selection of a hyperplane as an optimization problem for which there are effective algorithms. And, the problem of ensuring good generalization performance is addressed by searching for a hyperplane with large margin. These ideas taken together result in a general and highly successful learning method.

17.1 MAPPING THE FEATURE VECTORS

As mentioned above, the problem of representation ability is solved by first mapping the training examples to a high-dimensional space in a nonlinear way and then using linear rules in this new space. This results in nonlinear rules in the original feature space.

In a sense, we can think of multilayer networks as doing something similar. Consider Figure 10.10, repeated here as Figure 17.1. The single perceptron in the output layer performs a threshold on the weighted linear combination of the inputs to this perceptron. The inputs to this perceptron happen to be some nonlinear transformation of the original feature vector, even though it is not clear exactly what nonlinear transformation is performed. In contrast, we could directly specify a nonlinear transformation and then find a linear classifier in the transformed space.

For a simple example, consider the XOR problem in two dimensions discussed in Chapter 9. Recall that those points $\overline{x} = (x_1, x_2)$ for which $x_1 x_2 > 0$ are labeled 1 and those points for which $x_1 x_2 < 0$ are labeled -1. See Figure 9.5, repeated here as Figure 17.2. Linear rules are clearly inadequate for this problem. However, consider transforming the vector $\overline{x}$ to a new three-dimensional vector $\overline{z} = (z_1, z_2, z_3)$ such that $z_1 = x_1, z_2 = x_2$, and $z_3 = x_1 x_2$ (Figure 17.3). Then, the linear rules in the transformed space are adequate to represent the original XOR problem since in the new space the simple (hyper)plane $z_3 = 0$ separates points labeled 1 from those points labeled -1. In this example, we have increased the dimension by 1, going from $\mathbf{R}^2$ to $\mathbf{R}^3$. In this particular case to get separation we could have actually *decreased* the dimension by 1, by simply mapping $\overline{x}$ to z_3. However, this is a result of the special structure of the XOR problem and the fact that we know this structure ahead of time.

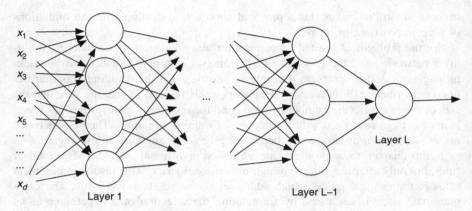

Figure 17.1 .Feed forward neural network.

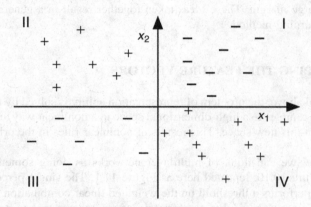

Figure 17.2   XOR representation problem for perceptrons.

Usually, the problem may not have such a simple structure and we may know very little about whatever structure is inherent in the problem. The mapping will typically be more involved and will transform $\overline{x}$ into a much higher dimensional space in order to provide a rich set of possible rules in the original space. The new feature space can even be infinite dimensional! For example, with only two dimensions in the original space for x_1 and x_2, the features in the transformed space might be $z_1 = x_1, z_2 = x_2, z_3 = x_1^2, z_4 = x_1 x_2, z_5 = x_2{}^2, z_6 = x_1^3, z_7 = x_1^2 x_2, \dots$.

Let $\mathcal{H}$ denote the new feature space, and let Φ denote the mapping, so that

$$\Phi : \mathbf{R}^d \to \mathcal{H}.$$

For an original feature vector $\overline{x} \in \mathbf{R}^d$, the transformed feature vector is given by $\Phi(\overline{x})$. The label y remains the same. Thus, the training example $(\overline{x}_i, y_i)$ becomes $(\Phi(\overline{x}_i), y_i)$.

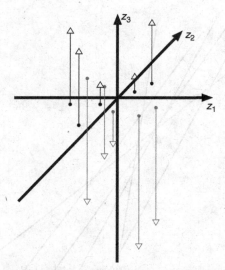

Figure 17.3 Mapping 2D XOR into 3D. The arrows point to where points in the (x_1, x_2) plane (which is the same as the (z_1, z_2) plane) are mapped to 3-space.

Then, we seek a hyperplane in the transformed space $\mathcal{H}$ that separates the transformed training examples $(\Phi(\overline{x}_1), y_1), \ldots, (\Phi(\overline{x}_n), y_n)$. That is, we want to find a hyperplane in the space $\mathcal{H}$ so that a transformed feature vector $\Phi(\overline{x}_i)$ lies on one side of the hyperplane if the label $y_i = -1$ and $\Phi(\overline{x}_i)$ lies on the other side of the hyperplane if $y_i = 1$. As in Chapters 9 and 10, it is convenient here to assume that the class labels are -1 and 1 instead of 0 and 1.

17.2 MAXIMIZING THE MARGIN

In general, we do not expect that the training data $(\overline{x}_1, y_1), \ldots, (\overline{x}_n, y_n)$ will be linearly separable. But, if transformed into a sufficiently high-dimensional space, the transformed data $(\Phi(\overline{x}_1), y_1), \ldots, (\Phi(\overline{x}_n), y_n)$ might be. In any case, for the moment let us assume that the data are separable by a hyperplane. Then there are generally infinitely many such separating hyperplanes. For example, Figure 17.4 shows a separable case in two dimensions with a number of separating lines.

All of these separating hyperplanes perform equally well on the training data (in fact they perform perfectly), but it is not clear how they will perform on new examples. That is, each of the separating hyperplanes might have different generalization performance. It is natural to ask whether some of the separating hyperplanes are better than others in terms of their error rate on new examples.

Without some qualification, the answer is no. From VC theory, it can be shown that any linear classifier that separates the data in d dimensions will have an expected error rate less than some fixed constant times $d \log n / n$. Although this

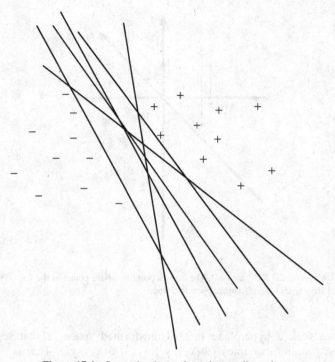

Figure 17.4 Separating hyperplanes in two dimensions.

bound may be acceptable if the dimension d is small compared with the amount of training data n, in some applications d may be very large, in which case the bound can be quite poor. In fact, if we map the training examples as discussed in the previous section, then d actually represents the dimension of the transformed space $\mathcal{H}$ so that we are explicitly making d large (possibly even infinite), which results in an unsatisfactory bound.

It can also be shown that this bound is tight in the sense that for any classifier, there is a distribution for which the error rate is close to the bound. Thus, in a worst-case sense it does not matter which separating hyperplane is selected—all will result in similar worst-case performance.

However, it may be that the distributions giving rise to the worst-case performance are unusual, and that for most "typical" distributions we might be able to do better by exploiting some properties of the training data. A key idea is to try to find hyperplanes that separate the data with a *large margin*. That is, given a classifier that separates the data, we can measure the distance from the training examples to the boundary of the classifier. In particular, let d_+ denote the smallest distance from examples labeled 1 to the separating hyperplane, and let d_- denote the smallest distance from examples labeled -1 to the separating hyperplane. The margin of the hyperplane is defined to be $d_+ + d_-$. By choosing the right orientation of a

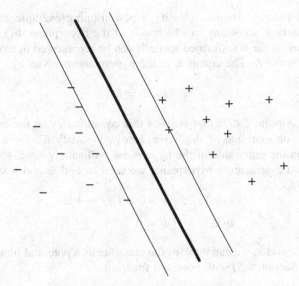

Figure 17.5 Large-margin separation.

separating hyperplane, we can make $d_+ + d_-$ as large as possible. Then any plane parallel to this will have the same value of $d_+ + d_-$.

In Figure 17.5, we show this in two dimensions. The hyperplane for which $d_+ = d_-$ (i.e., that is equally distant from the positive and negative examples) is shown, together with parallel hyperplanes for which either $d_+ = 0$ or $d_- = 0$. These last two hyperplanes will pass through one or more examples from the training data. These examples are called the *support vectors*, and they are the examples that define the maximum margin hyperplane. If we move or remove the support vectors, then the maximum margin hyperplane can change, but if we move or remove any of the other training examples, then the maximum margin hyperplane remains unchanged.

Intuitively, if the margin is large then the separation of the training examples is robust to small changes in the hyperplane, and we would expect the classifier to have better predictive performance. This is often the case, though not in a worst-case sense. The fact that large margin classifiers tend to have good generalization performance has been both justified theoretically and observed in practice.

17.3 OPTIMIZATION AND SUPPORT VECTORS

From our discussion so far, SVMs roughly operate as follows. The original training examples are $(\overline{x}_1, y_1), \ldots, (\overline{x}_n, y_n)$, where the $\overline{x}_i \in \mathbf{R}^d$ and $y_i \in \{-1, 1\}$. These training examples are transformed to a high-dimensional space $\mathcal{H}$ using a mapping Φ to produce $(\Phi(\overline{x}_1), y_1), \ldots, (\Phi(\overline{x}_n), y_n)$, where $\Phi(\overline{x}_i) \in \mathcal{H}$. In the transformed space $\mathcal{H}$ we seek a hyperplane that separates the transformed training examples with maximal margin. Then to classify a new training example $\overline{x}$, we compute $\Phi(\overline{x})$ and classify $\overline{x}$ according to which side of the hyperplane $\Phi(\overline{x})$ lies.

A hyperplane in the transformed space $\mathcal{H}$ can be represented in terms of a vector $\overline{w} \in \mathcal{H}$ and a scalar b. The equation of the hyperplane is given by

$$\overline{w} \cdot \overline{z} + b = 0,$$

where $\overline{z} \in \mathcal{H}$. A point $\overline{z} \in \mathcal{H}$ that satisfies this equation lies on the hyperplane. All points that lie on one side of the hyperplane will satisfy $\overline{w} \cdot \overline{z} + b > 0$ while all points lying on the other side of the hyperplane will satisfy $\overline{w} \cdot \overline{z} + b < 0$.

Thus, to find a separating hyperplane, we need to find $\overline{w}$ and b such that

$$\Phi(\overline{x}_i) \cdot \overline{w} + b > 0 \quad \text{if } y_i = +1,$$

$$\Phi(\overline{x}_i) \cdot \overline{w} + b < 0 \quad \text{if } y_i = -1.$$

Given such a $\overline{w}$ and b, we can think of the classifier as a potential function classifier (discussed in Section 8.5) with potential function

$$f(\overline{x}) = \overline{w} \cdot \Phi(\overline{x}) + b. \tag{17.1}$$

Among all separating hyperplanes we would like one with maximum margin. This task can be formulated as an optimization problem. Actually, even after applying the mapping, the transformed examples $(\Phi(\overline{x}_i), y_i)$ might still not be separable by a hyperplane. In this case, the optimization problem that we solve balances separating the data as much as possible while also trying to maximize the margin.

Through solving the optimization problem, we obtain $\overline{w}$ and b that define the maximum margin hyperplane. It turns out that $\overline{w}$ obtained in this way is given by

$$\overline{w} = \sum_{i=1}^{n} \alpha_i y_i \Phi(\overline{x}_i), \tag{17.2}$$

where the α_i are nonnegative numbers that are determined by solving the optimization problem. The constant b is also obtained through solving the optimization problem.

For some i, it turns out that $\alpha_i = 0$. For these i, the corresponding example $(\Phi(\overline{x}_i), y_i)$ does not affect the maximum margin hyperplane. For other i, we have $\alpha_i > 0$, and the examples corresponding to these i do affect the maximum margin hyperplane. These examples (corresponding to positive α_i) are the *support vectors*.

Substituting the expression for $\overline{w}$ given in Equation (17.2) into the expression for the potential function in Equation (17.1), we get

$$\overline{w} \cdot \Phi(\overline{x}) + b = \sum_{i=1}^{n} \alpha_i y_i (\Phi(\overline{x}_i) \cdot \Phi(\overline{x})) + b. \tag{17.3}$$

17.4 IMPLEMENTATION AND CONNECTION TO KERNEL METHODS

An important practical consideration is how to solve the optimization problem and actually construct such a maximum margin hyperplane in the transformed space. One part of the problem is how to implement the transformation $\Phi(\cdot)$. The original feature vector $\overline{x}$ is often high dimensional itself and the transformed space is typically of even much higher dimension, possibly even infinite dimensional. Thus, computing $\Phi(\overline{x}_i)$ and $\Phi(\overline{x})$ can be very difficult.

A useful result from mathematics allows us to address this issue and also provides a connection between SVMs and kernel methods discussed in Chapter 8. It turns out (as we discuss in Section 17.5) that the optimization problem can be written in terms of dot products of the form $\Phi(\overline{x}_i) \cdot \Phi(\overline{x}_j)$. Also, as seen from Equation (17.3), implementing the corresponding decision rule to classify a feature vector $\overline{x}$ involves dot products of the form $\Phi(\overline{x}_i) \cdot \Phi(\overline{x})$.

Under certain conditions the dot product $\Phi(\overline{x}_i) \cdot \Phi(\overline{x})$ can be replaced with a function $K(\overline{x}_i, \overline{x})$ that is easy to compute. This function $K(\cdot, \cdot)$ is just a kernel function, and the resulting classifier becomes a form of a general kernel classifier, that is,

$$\overline{w} \cdot \Phi(\overline{x}) + b = \sum_{i=1}^{n} \alpha_i y_i \left(\Phi(\overline{x}_i) \cdot \Phi(\overline{x}) \right) + b$$

$$= \sum_{i=1}^{n} \alpha_i y_i K(\overline{x}_i, \overline{x}) + b.$$

The terms in the optimization problem used to find the α_i also contain dot products of the form $\Phi(\overline{x}_i) \cdot \Phi(\overline{x}_j)$ which can be replaced with $K(\overline{x}_i, \overline{x}_j)$. So, a feature vector $\overline{x}$ is classified as 1 if $\sum_i \alpha_i y_i K(\overline{x}_i, \overline{x}) + b > 0$ and $\overline{x}$ is classified as -1 if $\sum_i \alpha_i y_i K(\overline{x}_i, \overline{x}) + b < 0$.

In practice, one generally directly chooses the kernel function K, while the mapping $\Phi(\cdot)$ and the transformed space $\mathcal{H}$ are induced by the choice of K. In fact, once we specify a kernel, the training and classification rule can be implemented directly without the need for even knowing what are the corresponding Φ and $\mathcal{H}$. It turns out that for a given choice of the kernel K, corresponding Φ and $\mathcal{H}$ exist if and only if K satisfies a condition known as *Mercer's condition* — that is, for all $g(\overline{x})$ such that $\int g(\overline{x})^2 \, d\overline{x} < \infty$, we have $\int K(\overline{x}, \overline{z}) g(\overline{x}) g(\overline{z}) \, d\overline{x} \, d\overline{z} \geq 0$.

Some common choices of kernel functions are as discussed in Chapter 8, namely:

$$K(\overline{x}_i, \overline{x}_j) = (\overline{x}_i \cdot \overline{x}_j + 1)^p,$$

$$K(\overline{x}_i, \overline{x}_j) = e^{-\|\overline{x}_i - \overline{x}_j\|^2 / 2\sigma^2},$$

$$K(\overline{x}_i, \overline{x}_j) = \tanh(\kappa(\overline{x}_i \cdot \overline{x}_j) - \delta).$$

As with kernel methods in general, the choice of the kernel function and associated parameters is an art and can have a strong influence on classification performance.

17.5   DETAILS OF THE OPTIMIZATION PROBLEM*

In this section, we give the details of the optimization problem for finding the maximum margin hyperplane. We first rewrite the separation equation and give the precise equation for the margin when the data are separated by a hyperplane. We then introduce the idea of slack variables in case the training examples are *not* linearly separable even in the transformed space. This requires an appropriate modification of the notion of "maximum margin." Finally, we describe how the optimization problem is reformulated using tools from optimization theory and give the equations for the solution.

17.5.1   Rewriting Separation Conditions

Recall that the transformed training examples are $(\Phi(\overline{x}_1), y_1), \ldots, (\Phi(\overline{x}_n), y_n)$, where $\Phi(\overline{x}_i) \in \mathcal{H}$ and $y_i \in \{-1, 1\}$. The equation of a hyperplane in $\mathcal{H}$ can be represented in terms of a vector $\overline{w}$ and a scalar b as

$$\overline{w} \cdot \Phi(\overline{x}) + b = 0.$$

It turns out that $\overline{w}$ is the normal to the hyperplane and $|b|/\|\overline{w}\|$ is the distance of the hyperplane from the origin.

If the hyperplane separates the training data, then

$$\Phi(\overline{x}_i) \cdot \overline{w} + b > 0 \ \text{ if } y_i = +1,$$

$$\Phi(\overline{x}_i) \cdot \overline{w} + b < 0 \ \text{ if } y_i = -1.$$

These equations simply formalize the requirement that those $\Phi(\overline{x}_i)$ that are labeled $+1$ lie on one side of the hyperplane while those $\Phi(\overline{x}_i)$ labeled -1 lie on the other side. Since there are only finitely many $\Phi(\overline{x}_i)$ and they are all some nonzero distance from the hyperplane, we have

$$\Phi(\overline{x}_i) \cdot \overline{w} + b \geq \beta \ \text{ if } y_i = +1,$$

$$\Phi(\overline{x}_i) \cdot \overline{w} + b \leq -\beta \ \text{ if } y_i = -1,$$

for some $\beta > 0$. We can then renormalize the equations above (i.e., divide both sides by β and relabel $\overline{w}/\beta$ and b/β with just $\overline{w}$ and b) to achieve that a separating hyperplane satisfies

$$\Phi(\overline{x}_i) \cdot \overline{w} + b \geq +1 \ \text{ if } y_i = +1,$$

$$\Phi(\overline{x}_i) \cdot \overline{w} + b \leq -1 \ \text{ if } y_i = -1.$$

Equivalently, we can write these conditions as

$$y_i(\Phi(\overline{x}_i) \cdot \overline{w} + b) - 1 \geq 0 \quad \text{for } i = 1, \ldots, n. \tag{17.4}$$

17.5.2   Equation for Margin

With the separation condition written as in Equation (17.4), it turns out that the shortest distance from a positive example to the hyperplane (d_+) and the shortest distance from a negative example to the hyperplane (d_-) are both equal to $1/\|\overline{w}\|$ (Figure 17.5). The margin for the hyperplane is then given by

$$\text{margin} = d_+ + d_- = \frac{2}{\|\overline{w}\|}.$$

To maximize the margin we can minimize $\|\overline{w}\|$ or, equivalently, $\|\overline{w}\|^2$. (To find algorithms for actually carrying out optimizations, it is much easier to minimize the square of the magnitude of $\overline{w}$ rather than the magnitude itself.) Thus, we can find the maximum margin hyperplane that separates the training data, by solving the following optimization problem:

$$\text{minimize } \|\overline{w}\|^2$$
$$\text{subject to } y_i(\Phi(\overline{x}_i) \cdot \overline{w} + b) - 1 \geq 0 \quad \text{for } i = 1, \ldots, n.$$

17.5.3   Slack Variables for Nonseparable Examples

In general, we may not be able to separate the training data with a hyperplane. However, we can still seek a hyperplane that separates the data "as much as possible" while also trying to "maximize the margin," with these objectives suitably defined.

In order to carry this out, we introduce "slack variables" ξ_i with $\xi_i \geq 0$ for $i = 1, \ldots, n$ and try to satisfy

$$\Phi(\overline{x}_i) \cdot \overline{w} + b \geq +1 - \xi_i \text{ if } y_i = +1,$$
$$\Phi(\overline{x}_i) \cdot \overline{w} + b \leq -1 + \xi_i \text{ if } y_i = -1.$$

Since the ξ_i are nonnegative, we see that satisfying the conditions above is a relaxed version of the original separation condition. The ξ_i allow these conditions to be satisfied with a certain amount of "slack." Without a constraint on the ξ_i, the conditions above can be satisfied trivially. That is to say, one could take *any* hyperplane and then select ξ_i large enough so that the conditions are satisfied.

We can get a well-posed problem by adding a penalty term that includes the slack variables to our optimization problem. In particular, since we have already constrained the ξ to be nonnegative, we get a useful formulation by adding a term of the form $C \sum_i \xi_i$, where C is some appropriate constant. With this term, larger ξ_i are penalized, and this prevents the degenerate satisfaction of the conditions by taking the ξ_i to be arbitrarily large. Thus, we seek a hyperplane that solves the following optimization problem:

$$\text{minimize } \|\overline{w}\|^2 + C \sum_i \xi_i$$
$$\text{subject to } y_i(\Phi(\overline{x}_i) \cdot \overline{w} + b) - 1 + \xi_i \geq 0 \quad \text{for } i = 1, \ldots, n$$
$$\xi_i \geq 0 \quad \text{for } i = 1, \ldots, n.$$

17.5.4 Reformulation and Solution of Optimization

Using techniques from optimization (Lagrange multipliers and considering the dual problem), the maximum margin separating hyperplane can be found by solving the following optimization problem:

$$\text{maximize } \sum_i \alpha_i - \frac{1}{2} \sum_{i,j} \alpha_i \alpha_j \, y_i y_j \, (\Phi(\overline{x}_i) \cdot \Phi(\overline{x}_j))$$
$$\text{subject to } \sum_i \alpha_i y_i = 0$$
$$0 \leq \alpha_i \leq C \quad \text{for} \quad i = 1, \ldots, n.$$

This is a standard type of optimization problem called a *convex quadratic programming problem* for which there are well-known and efficient algorithms. Through solving this optimization, the following equations are satisfied, which provide the equation for the separating hyperplane:

$$\overline{w} = \sum_{i=1}^{n} \alpha_i y_i \Phi(\overline{x}_i), \tag{17.5}$$

$$\alpha_i(y_i(\overline{w} \cdot \Phi(\overline{x}_i) + b) - 1 + \xi_i) = 0 \text{ for } i = 1, \ldots, n, \tag{17.6}$$

$$y_i(\Phi(\overline{x}_i) \cdot \overline{w} + b) - 1 + \xi_i \geq 0 \text{ for } i = 1, \ldots, n. \tag{17.7}$$

The solution to the optimization problem returns values for the α_i and the ξ_i. From these, $\overline{w}$ can be obtained directly from the expression above in Equation (17.5). The scalar b can be obtained by solving Equation (17.6) for any i in terms of b, though a better approach is to do this for all i and take the average of the values obtained as the choice for b.

Solving this optimization problem and obtaining the separating hyperplane (namely, $\overline{w}$ and b) are the processes of training. For classification, we need to only check on which side of the hyperplane a feature vector $\overline{x}$ falls. That is, $\overline{x}$ is classified as 1 if

$$\overline{w} \cdot \Phi(\overline{x}) + b = \sum_{i=1}^{n} \alpha_i y_i \left(\Phi(\overline{x}_i) \cdot \Phi(\overline{x}) \right) + b > 0,$$

and $\overline{x}$ is classified as -1 otherwise.

As mentioned in Section 17.4, under certain conditions the dot product $\Phi(\overline{x}_i) \cdot \Phi(\overline{x}_j)$ can be replaced with a function $K(x_i, x_j)$ that is easy to compute. The resulting classifier then becomes a form of a general kernel classifier. That is, a feature vector $\overline{x}$ is classified according to whether

$$\overline{w} \cdot \Phi(\overline{x}) + b = \sum_{i=1}^{n} \alpha_i y_i \left(\Phi(\overline{x}_i) \cdot \Phi(\overline{x}) \right) + b$$

$$= \sum_{i=1}^{n} \alpha_i y_i \, K(\overline{x}_i, \overline{x}) + b$$

is greater than zero or less than zero.

17.6 SUMMARY

In this chapter we described a highly successful state-of-the-art learning approach known as SVMs. We started by discussing a key idea of SVMs which is to map the feature vectors to a high-dimensional space and find a separating hyperplane in this transformed space. This allows nonlinear classifiers in the original space, and thus allows us to overcome the representational limitations of linear classifiers. Yet, by working with linear classifiers we can come up with efficient computational methods to find a classifier.

A second key idea of SVMs is that of large-margin classifiers. Among all hyperplanes that separate a given set of training data, those with large margin will have significantly better classification performance for typical distributions than others, even though in a worst-case sense all separating hyperplanes have comparable performance. Thus, a large-margin classifier in the transformed space can perform well, even though the transformed space has high dimension.

This approach of transforming the feature space and finding a large-margin classifier can be formulated as an optimization problem. Under certain conditions, the classifier resulting from the solution to the optimization problem takes the form of kernel methods and can be efficiently implemented.

The optimization problem is most easily formulated in the case where the nonlinear mapping is not used and the original training data are linearly separable. In the case when the training data are not separable, slack variables can be introduced and an optimization problem can be formulated that attempts to find a hyperplane that separates the data as much as possible, that is, has a margin as large as possible and minimizes the slack variables. The general case of nonlinear classifiers (i.e., using the mapping to a higher dimensional space) and nonseparable data in the transformed case results in the kernel methods mentioned above. Often in implementations of SVMs, the kernels are selected and the choice of kernel induces a mapping and a new high-dimensional space, even though these need not be explicitly considered.

17.7 APPENDIX: COMPUTATION

The success of SVMs as a practical learning method is due to a variety of factors, one of which is the availability of effective algorithms for solving the corresponding optimization problems. This is true more generally for successful learning methods. That is, in addition to the generalization performance and other attributes of a learning method, having good algorithms is crucial.

One might ask when good algorithms exist for learning or other problems and how does one even formalize such a question? And before asking about the existence of a good algorithm, one needs to address the question of what is an algorithm? A branch of computer science known as the *theory of computation* addresses these and other questions.

The modern theory of computing is often considered to have started with the work of Alan Turing, Alonzo Church, and others. Turing proposed a fairly simple and intuitive mathematical model for computing known as a *Turing machine* that tries to capture the essence of digital computation. Some other models such as recursion and λ-calculus being considered around the same time were shown to have the same computational power as Turing machines. Since then a number of other models for computing (including many variations in the details of the Turing machine model) also have been shown to have the same computational power. Turing machines (or other equivalent models) formalize the notion of an algorithm and what can be computed. The equivalence of many models gives a reassuring sense of robustness in the definition and leads to the widely accepted hypothesis that Turing machines capture the right notion of computing—that is, anything that is computable can be computed by a Turing machine, which is often known as *Church's thesis* or the *Church–Turing thesis*.

A remarkable result is that there are some problems or functions (in fact many problems/functions) that cannot be solved or computed by a Turing machine, and hence by any of the other equivalent models of computation. For such problems/functions, it is not a question of whether there is a good algorithm to solve the problem or compute the function. Such problems simply cannot be solved and such functions simply cannot be computed!

But fortunately many (all?) problems of interest are computable! For such problems, algorithms do exist for solving the problem and so some natural questions are whether there are good algorithms and what is meant by good. Such questions are studied in a branch of computer science called *complexity theory*.

A very common approach in complexity theory is to consider the time (number of steps) and space (amount of memory) needed to solve a problem. More specifically, this type of analysis usually looks at how the time and/or space required grow with the size of the problem. For example, in the case of learning, the size of the problem might be captured by the dimension of the feature space d and the number of training examples n. If we fix the dimension d, we might be interested in how long it takes to compute a good decision rule as a function of the number of training examples. We usually look for algorithms where the time required does not grow too rapidly as a function of n.

Imagine, for example, three algorithms. The first requires $0.4n$ s, the second algorithm requires $0.03n^2$ s, and the third requires $(0.001)2^n$ s to compute a good rule. If we have 10 training examples, then the 3 algorithms would require 4, 3, and 1.024 s, respectively. It looks like the third algorithm is quite fast. But, if we had 1000 training examples, then the first algorithm takes 400 s, the second algorithm takes 30,000 s, and the third algorithm takes over 10^{297} s! Clearly, the efficiency of an algorithm is an important consideration for practical applications.

17.8 QUESTIONS

1. True or False: In SVMs, using linear rules in the transformed space can provide nonlinear rules in the original feature space.

2. What is the form of the decision rule for a linear SVM?

3. If we insist that the hyperplane pass through the origin, what will the form of the rule be?

4. Show that by increasing the dimension of the feature vector by 1, a hyperplane passing through the origin in the augmented space can represent a general hyperplane decision rule in the original space.

5. How does the construction of the previous problem affect the margin of the classifier?

6. SVMs solve a classification problem by, in effect, transforming the feature space into another typically higher dimensional, sometimes infinitely dimensional space and then finding a (possibly soft) wide margin linear separation in that other space. What is the VC-dimension of linear classifications in an infinitely dimensional space? Is that a problem? If so, does the use of wide margin separations address that problem? Explain.

7. Why are SVMs called *support vector* machines. What is a support vector and what role does it play in an SVM?

8. Suppose that in the transformed space the feature vectors and corresponding labels are given as follows:

$$(-1, 0; +), (0, 1; +), (1, 0; +), (-1, 2; -), (-1, 3; -), (0, 3; -), (1, 5; -).$$

(a) Plot the feature vectors. Label the support vectors and solve for the margin. What is the SVM decision rule in this case? How would it classify the feature vector (1,2)?

(b) Now we augment our data by adding a new point (−1,1;+). Repeat part (a).

(c) If there were another data point in the set (0,0;−), describe in broad terms how we would have to change our SVM algorithm. Why is this so?

17.9 REFERENCES

SVMs have their origin in the work of Vapnik (1979), but did not attract wide attention until the early 1990s through the work of Vapnik and others, for example, Boser *et al.* (1991), Cortes and Vapnik (1995), and Vapnik (1991). Since then SVMs have been widely studied and have been successfully applied in a broad range of applications, for example, Poldrak *et al.* (2009).

The tutorial by Burges (1998) and the books by Cristianini and Shawe-Taylor (2000), Shawe-Taylor and Cristianini (2004), Schölkopf and Smola (2001), and Vapnik (1991, 1998) provide good entry points to SVMs. The edited books by Schölkopf *et al.* (1999) and Bartlett *et al.* (2000) contain collections of papers in these areas.

Bartlett P, Schölkopf B, Schuurmans D, Smola AJ, editors. Advances in large-margin classifiers. Cambridge (MA): MIT Press; 2000.

Boser B, Guyon I, Vapnik VN. A training algorithm for optimal margin classifiers. Proceedings of the 5th Annual ACM Workshop on Computational Learning Theory. New York: Association for Computing Machinery; 1992. pp. 144–152.

Burges CJC. A tutorial on support vector machines for pattern recognition. Data Mining Knowl Discov 1998;2:121–167.

Cortes C, Vapnik VN. Support vector networks. Mach Learn 1995;20:1–25.

Cristianini N, Shawe-Taylor J. An introduction to support vector machines. Cambridge: Cambridge University Press; 2000.

Poldrak RA, Halchenko YO, Hanson SJ. Decoding the large-scale structure of brain function by classifying mental states across individuals. Psychol Sci 2009;20(11):1364–1372.

Schölkopf B, Burges CJC, Smola AJ, editors. Advances in kernel methods: support vector learning. Cambridge (MA): MIT Press; 1999.

Schölkopf B, Smola AJ. Learning with kernels: support vector machines, regularization, optimization, and beyond (adaptive computation and machine learning). Cambridge (MA): MIT Press; 2001.

Shawe-Taylor J, Cristianini N. Kernel methods for pattern analysis. Cambridge: Cambridge University Press; 2004.

Vapnik VN. Esimation of Dependencies Based on Empirical Data. Moscow: Nauka, 1979, in Russian. English translation New York: Springer; 1982.

Vapnik VN. The nature of statistical learning theory. New York: Springer; 1991.

Vapnik V. Statistical learning theory. New York: Wiley-Interscience; 1998.

CHAPTER 18

Boosting

Boosting is an iterative procedure for improving the performance of any learning algorithm. It is among the most successful learning methods available. Boosting combines a set of "weak" classification rules to produce a "strong" composite classifier. Boosting proceeds in a series of rounds. In each round, a weak rule is produced by running some basic learning algorithm using a different weighting of the training examples. Starting with equal weighting in the first round, the weighting of the training examples is updated after each round to place more weight on those training examples that are misclassified by the current weak hypothesis just produced and less weight on those that are correctly classified. This forces the weak learner to concentrate on these hard-to-classify examples in the next round. After a number of rounds, a final classification rule is produced via a weighted sum of the weak rules. Although the weak rules produced in each round may have modest performance, the algorithm "boosts" the performance to produce a new composite rule with much better performance.

There are a number of variants of the basic boosting algorithm, one of the most popular of which is the AdaBoost algorithm developed by Freund and Schapire (1997). Before describing AdaBoost, we provide some background on weak learners and two of the key ingredients that are needed, namely, the ideas of combining classification rules and of updating the weightings of the training examples.

18.1 WEAK LEARNING RULES

Recall that in any classification problem, the best error rate we can ever hope to achieve is the Bayes error rate R^*. For any problem, it is trivial to design a learning rule that has error rate $1/2$. All we need to do is ignore the feature vector $\bar{x}$ and decide $+1$ or -1 randomly. Thus, as mentioned in Chapter 5, we certainly have $R^* \leq 1/2$.

An Elementary Introduction to Statistical Learning Theory, First Edition.
Sanjeev Kulkarni and Gilbert Harman.
© 2011 John Wiley & Sons, Inc. Published 2011 by John Wiley & Sons, Inc.

There are some problems for which $R^* = 1/2$. For these problems, we can do no better than random guessing, regardless of the learning algorithm. For such problems, boosting (or any other learning algorithm for that matter) will be of no help. However, more typically, the Bayes error rate is strictly less than 1/2. For such problems we would like to find rules with performance close to the Bayes error rate.

As we have seen, the usual procedure is to observe a set of training examples $(\overline{x}_1, y_1), \ldots, (\overline{x}_n, y_n)$ and to use the training data to construct classification rules. Finding a rule with performance close to the Bayes rate can be quite difficult (and we generally do not even know what is the Bayes rate). We have described a variety of techniques each of which has some advantages and disadvantages, depending on the problem at hand, and some of these will produce rules with error rate close to the Bayes rate in the limit of infinitely many training examples. This was the universal consistency property discussed earlier. However, for finite sample sizes, all bets are generally off unless we impose some conditions on the underlying distributions, the class of rules under consideration, or both.

Thus, for a finite number of training examples, the performance of the rule produced might be far from the best possible. However, if the rule is performing strictly better than random guessing, then it has extracted something useful from the training examples. Such a rule will be called a *weak* learning rule. A weak learning rule has an error rate $\epsilon = 1/2 - \gamma$ for some $\gamma > 0$.

Because of computational or other considerations, it might be easy to produce a weak learning rule, even though finding a good rule may be difficult. This is where boosting comes into play. If we have a method for producing even weak rules, boosting can be used to combine these weak rules to produce a new classifier with much better performance.

18.2 COMBINING CLASSIFIERS

Boosting combines a collection of weak classifiers to form a composite classifier with (hopefully) better performance. This new classification rule is formed from a weighted combination of the original classifiers. We explain how the weights are computed later, and in this section we describe how to combine the classifiers if we are given the weights.

Given a set of classification rules $h_1(\overline{x}), h_2(\overline{x}), \ldots, h_T(\overline{x})$, we can form a new classification rule as a weighted combination of these as follows. Suppose we have a set of weights $\alpha_1, \alpha_2, \ldots, \alpha_T$. Define

$$H(\overline{x}) = \text{sign}\left(\sum_{t=1}^{T} \alpha_t h_t(\overline{x})\right),$$

where as before (Chapter 9), $\text{sign}(u)$ returns -1 if $u < 0$ and returns 1 otherwise.

For a given $\overline{x}$, each individual rule $h_t(\overline{x})$ outputs either $+1$ or -1. The combined rule $H(\overline{x})$ is a sign of some quantity; namely, it is the sign of the weighted sum

$\sum_{t=1}^{T} \alpha_t h_t(\overline{x})$. If this weighted sum is positive, then $H(\overline{x}) = 1$. If the weighted sum is negative, then $H(\overline{x}) = -1$. Hence, $H(\overline{x})$ also outputs either $+1$ or -1, as desired. Of course, we need some way to handle the case when the weighted sum is zero. For simplicity, let us just assume that this gets mapped to $+1$. We already did this implicitly with the definition of the sign($\cdot$) function, as we did in Chapter 9. Then for any feature vector $\overline{x}$, the output is either $+1$ or -1 so that $H(\overline{x})$ is a classification rule in the usual sense.

Note that the individual output $h_t(\overline{x})$ is the decision made by the weak classifier produced in round t. Each of these decisions is multiplied by the corresponding weight α_t and these are added together to produce the weighted sum. In a special case in which the α_t are all equal and positive, $H(\overline{x})$ is just the majority vote of the individual rules $h_t(\overline{x})$. But, usually, the decisions of certain rules $h_t(\overline{x})$ will be weighted more heavily than others.

18.3 DISTRIBUTION ON THE TRAINING EXAMPLES

As usual, the training examples to be used for learning are $(\overline{x}_1, y_1), \ldots, (\overline{x}_n, y_n)$, where $\overline{x}_i \in \mathbf{R}^d$ and $y_i \in \{-1, +1\}$. (Again, in this chapter it is convenient to consider the classes as -1 and $+1$ instead of 0 and 1). Normally, any learning algorithm will try to find a classification rule that makes the training error small. Given a rule $h(\cdot)$, the training error of h is measured by the number of errors on the training examples, that is,

$$\sum_{i=1}^{n} I_{\{h(\overline{x}_i) \neq y_i\}} = \sum_{i:h(\overline{x}_i) \neq y_i} 1,$$

where (as in Chapter 8) I_A denotes the so-called indicator function of the event A for which $I_A = 1$ if the event A is true and $I_A = 0$ otherwise. This is just a count of the number of training examples for which the rule h applied to feature vector $\overline{x}_i$ disagrees with the observed label y_i.

If we divide by n, then instead of the number of training examples on which h disagrees, we get the fraction of training examples for which h disagrees with the observed label. Let us denote this fraction by ϵ, so that

$$\epsilon = \frac{1}{n} \sum_{i=1}^{n} I_{\{h(\overline{x}_i) \neq y_i\}} = \frac{1}{n} \sum_{i:h(\overline{x}_i) \neq y_i} 1. \qquad (18.1)$$

We can think of ϵ as the probability of error (or error rate) over the training examples when each of the training examples is equally likely. Learning algorithms typically attempt to find a rule to make ϵ small, though as we have seen, sometimes this objective is balanced with a competing objective of simplicity of the classification rule.

The examples that are misclassified by h are presumably "difficult" ones to classify. Suppose we apply whatever learning algorithm we are using and try to focus attention on these difficult cases. We might do this, for example, by assigning more weight to these examples and less weight to the "easy" ones. We can think of the weight of a particular example as the probability of seeing that example. We can either generate "new" examples by drawing $(\overline{x}_i, y_i)$ pairs according to these probabilities, and then train the learning algorithm on this set of "new" examples. Or, if possible, we can simply apply the learning algorithm with the objective of minimizing the weighted training error, rather than the unweighted error ϵ defined above.

We will let $D_t(\cdot)$ denote the distribution on the training examples at round t, so that $D_t(i)$ is the probability (or weight) that we assign to the ith training example. At stage t, the goal of the weak learning algorithm will be to minimize the *weighted* training error. That is, instead of trying to minimize the unweighted error in Equation (18.1), the weak learning algorithm will produce a hypothesis $h_t(\cdot)$ that attempts to minimize the weighted error, ϵ_t, defined by

$$\epsilon_t = \sum_{i=1}^{n} D_t(i) I_{\{h_t(\overline{x}_i) \neq y_i\}}. \tag{18.2}$$

We can think of ϵ_t as the probability of error of the classifier $h_t(\cdot)$ produced by the weak learning algorithm at round t, where the probability is computed with respect to the distribution $D_t(\cdot)$.

18.4 THE ADABOOST ALGORITHM

We now have all the ingredients needed to describe the specifications of a boosting algorithm, but before giving the mathematical details, we give a rough description of the main steps.

Boosting works in the following way. As input, we have the training data $(\overline{x}_1, y_1), \ldots, (\overline{x}_n, y_n)$ and we have access to a weak learning algorithm. The output is a classifier that is a weighted combination of weak classifiers produced at each of several rounds. The main steps are as follows:

- Start with a uniform weighing that places equal weight on the training examples, and set $t = 1$.
- At each stage $t = 1, \ldots, T$, run the weak learning algorithm to produce a hypothesis $h_t(\cdot)$.
- Compute ϵ_t (the error rate of $h_t(\cdot)$), and use this to compute the weights α_t.
- Update the distribution on the training examples from $D_t(\cdot)$ to $D_{t+1}(\cdot)$ to assign more weight to those training examples that $h_t(\cdot)$ misclassified and less weight to those examples that $h_t(\cdot)$ classified correctly. That way, in the next round (i.e., at stage $t + 1$), the weak learning algorithm will focus more on the examples that were misclassified in the previous stage.
- After finishing all T rounds, use the weights α_t to form the final composite classifier $H(\cdot)$ as a weighted combination of the weak learning rules $h_t(\cdot)$.

To describe a boosting procedure precisely, we need to specify only the choice of the weights α_t and the way the distribution $D_t(\cdot)$ is updated. The version of boosting that we describe is the AdaBoost algorithm developed by Freund and Schapire (1995). This has been found to be an extremely effective algorithm and has been widely studied.

AdaBoost

- **Input:**
  - Training data $(\overline{x}_1, y_1), \ldots, (\overline{x}_n, y_n)$.
  - A weak learning algorithm.
- **Initialization:**
  - Set $t = 1$.
  - Set $D_1(i) = 1/n$.
- **Main Procedure:** For $t = 1, \ldots, T$:
  - Use the weak learning algorithm on distribution D_t to get classifier $h_t(\cdot)$.
  - Let ϵ_t be the error rate of $h_t(\cdot)$ with respect to the distribution D_t. Set $\alpha_t = \frac{1}{2} \log \left(\frac{1-\epsilon_t}{\epsilon_t} \right)$.
  - Update the distribution as follows:

$$D_{t+1}(i) = \frac{D_t(i)e^{-\alpha_t y_i h_t(\overline{x}_i)}}{Z_t},$$

  where Z_t is a normalization factor to ensure that

$$\sum_{t=1}^{T} D_{t+1}(i) = 1$$

- **Output:** The final classifier $H(\cdot)$ output by the boosting procedure is given by

$$H(\overline{x}) = \text{sign} \left(\sum_{t=1}^{T} \alpha_t h_t(\overline{x}) \right).$$

18.5 PERFORMANCE ON TRAINING DATA

Boosting is specifically designed to combine weak rules in such a way that the error on the training data can be reduced. Recall that we let ϵ_t denote the error rate of the hypothesis h_t produced at stage t weighted according to the distribution D_t. We have assumed that the base learner produces a hypothesis that is better than random guessing, so that $\epsilon_t < 1/2$. However, since the base algorithm is just a weak learner, the error rate may not be much better than random guessing. Let γ_t be the amount by which h_t performs better than random guessing on the training data (weighted according to D_t). That is,

$$\gamma_t = 1/2 - \epsilon_t.$$

It can be shown that the training error of the final classifier produced by AdaBoost is bounded as follows:

$$\frac{1}{n} \sum_{i=1}^{n} I_{\{H(\overline{x}_i) \neq y_i\}} \leq e^{-2 \sum_{t=1}^{T} \gamma_t^2}. \tag{18.3}$$

The left-hand side is just the (unweighted) error rate of H on the training data, that is, the number of training examples which H classifies incorrectly divided by the total number of training examples. Since the base classifier produced on each round is better than random guessing, $\gamma_t > 0$. Thus, the bound on the right-hand side of Equation (18.3) becomes smaller with every round.

Suppose the base classifier on each round is better than random guessing by some fixed amount. That is, suppose that for some $\gamma > 0$, we have that

$$\gamma_t \geq \gamma.$$

Then, the bound in Equation (18.3) shows that the training error goes to zero as the number of rounds increases. In fact, in this case the training error approaches zero exponentially fast in the number of rounds T since

$$e^{-2 \sum_{t=1}^{T} \gamma_t^2} \leq e^{-2 \sum_{t=1}^{T} \gamma^2} = e^{-2T\gamma^2}.$$

Boosting algorithms prior to AdaBoost satisfied similar bounds on the training error but required knowledge of γ, which can be hard to obtain. The beauty of AdaBoost is that it adapts naturally to the error rates of the h_t so that knowledge of γ is not required (hence the name AdaBoost, short for Adaptive Boosting).

18.6   GENERALIZATION PERFORMANCE

While the results in the previous section on the performance of boosting (in particular, AdaBoost) on the training data in some sense are reassuring, for a learning method to be valuable we are really interested in the performance on new data. That is, ultimately we are interested in the generalization error as opposed to the training error.

Of course, using VC theory, we have seen that if the set of decision rules under consideration is not too rich (as measured by the VC-dimension of this set), then we can bound the generalization performance for classifiers that minimize the training error. Such an approach can be applied to boosting with the following result. Let

V denote the VC-dimension of the set of base classifiers. As before, n denotes the number of training examples, and T denotes the number of rounds of boosting. Also, following the notation from previous chapters, the error rate of the classifier H on a new example is denoted by $R(H)$, and this quantifies the generalization performance. That is,

$$R(H) = P\{H(\overline{x}) \neq y\}$$

is the probability that H misclassifies a new randomly drawn example. After T rounds of boosting, the error rate is bounded, with high probability, as follows:

$$R(H) \leq \frac{1}{n} \sum_{i=1}^{n} I_{\{H(\overline{x}_i) \neq y_i\}} + O\left(\sqrt{\frac{TV}{n}}\right). \tag{18.4}$$

The first term on the right-hand side of Equation (18.4) is just the fraction of errors made by the classifier H on the training data as we saw before. From the previous section, we know that this tends to zero as the number of rounds increases. On the other hand, the second term on the right-hand side of Equation (18.4) is problematic. As T increases, this term also increases, so that the bound on the generalization error gets worse.

This suggests that as we run boosting for more rounds we will tend to overfit the training data. As the error on the training data becomes smaller, the bound on the generalization error gets worse, which is not unexpected. Although the base class has finite VC-dimension, boosting produces an output formed by a combination of a number of base classifiers. As T increases, the VC-dimension of the set of all such combinations generally increases.

However, although sometimes overfitting is observed, in many cases, in practice, the generalization error continues to decrease as the number of rounds increases, and this can happen even after the error on the training data is zero. Explanations for this have been provided in terms of the margins of the classifier. In Chapter 17, we introduced the idea of the margin of a linear classifier as a measure of how widely the classifier separates the data. There are different ways to measure this, and for the analysis of boosting a slightly different definition is used than that from Chapter 17.

Define the margin of the example $(\overline{x}_i, y_i)$ as

$$\text{margin}(\overline{x}_i, y_i) = \frac{y_i \sum_{t=1}^{T} \alpha_t h_t(\overline{x}_i)}{\sum_{t=1}^{T} \alpha_t}. \tag{18.5}$$

Recall that the classifier produced by boosting is

$$H(\overline{x}) = \text{sign}\left(\sum_{t=1}^{T} \alpha_t h_t(\overline{x})\right).$$

The quantity inside $\text{sign}(\cdot)$ is multiplied by y_i in the numerator of the expression for the margin. Hence, the numerator (and therefore the margin) of $(\overline{x}_i, y_i)$ is positive if and only if H classifies $(\overline{x}_i, y_i)$ correctly. Also, the margin is always between -1 and $+1$. It will be $+1$ if every one of the base classifiers h_t classifies $(\overline{x}_i, y_i)$ correctly and it will be -1 if every one of the base classifiers classifies $(\overline{x}_i, y_i)$ incorrectly.

In a sense, the margin measures the confidence in our classification of $(\overline{x}_i, y_i)$. If the margin of many of the training examples is large then H classifies these with high confidence, and we might expect that H also generalizes well. It has been shown that larger margins on the training example do in fact result in a better bound on the generalization performance of the classifier. In particular, it can be shown that for any $\theta > 0$, the error rate $R(H)$ is bounded, with high probability, by

$$R(H) \le \frac{1}{n} \sum_{i=1}^{n} I_{\{\mathrm{margin}(\overline{x}_i, y_i) \le \theta\}} + O\left(\sqrt{\frac{V}{n\theta^2}}\right). \qquad (18.6)$$

The first term of the bound is the fraction of training examples that have margin less than or equal to θ. An important feature of this bound is that it does not depend on T. This helps explain the empirical observation that often running boosting for many rounds does not increase the generalization error.

18.7 SUMMARY

Boosting is a very effective learning procedure that combines a set of weak classifiers to form a strong classifier. Boosting works through a series of rounds, and in each round a weak classifier is produced that focuses more attention on those training examples that have been misclassified in the previous round. The final classifier is then formed as a weighted combination of the weak classifiers produced in each round, where the weights are based on the performance of the weak classifiers. AdaBoost is a particular implementation of boosting that is simple, works well in applications, and has a number of useful properties.

18.8 APPENDIX: ENSEMBLE METHODS

We have considered a number of different methods for learning in this book, and there are many that we have not discussed. Faced with a specific learning problem, a natural question is which of the many available methods should we choose. As we discussed in Chapter 6 in the context of the so-called "no free lunch theorems," there is no algorithm that outperforms all others for all problems. Thus, it is important to make a judicious choice of the method on the basis of an understanding of the problem at hand. Once a method is chosen and applied to the training data, we obtain a decision rule.

The idea used in boosting is suggestive of a somewhat different approach. Rather than come up with a particular decision rule by simply applying a learning method to the training data, suppose we generate a number of decision rules, and then to classify a new feature vector we take a weighted majority vote among these rules. Boosting suggests one way to do this, and such approaches more generally have come to be called *ensemble methods*. The hope is that by using a collection (or ensemble) of classifiers and combining them in a suitable way, we can come up with a rule with a performance better than we could have obtained using any one of the individual classifiers.

There are many ways to generate an ensemble of classifiers that we can try to combine. For example, we could apply a number of different methods on the training data (e.g., the various methods we discussed as well as others), and the ensemble would consist of the collection of classifiers produced by the various methods. But for a practical approach, we really would like a systematic and computationally efficient way to generate the ensemble and combine the constituent classifiers. There are various ways to do this. Generally, a basic learning method is selected and then some details of the method or the way in which the training examples are used is systematically varied.

For example, if there are parameters or initial conditions used in the learning method, then the method can be run multiple times with different choices of the parameters or initial conditions. This has been used with neural networks where the network is trained multiple times using different random choices of initial weights.

Several ways to generate an ensemble of classifiers involve manipulating the training data. A technique known as *bagging* is a simple example of this approach. In bagging, if we have n training examples, then we generate a new set of n training examples by randomly selecting examples with replacement from the original set. By doing this, many examples will appear multiple times while others will be left out. The learning method is applied to this new set of training examples to produce a classifier and the process is repeated a number of times to produce the ensemble. This is a type of bootstrap technique and the term bagging comes from bootstrap aggregation. Another approach is to leave out disjoint subsets of the original training data and run the learning method on each reduced set of training examples. This is a type of cross-validation approach. Boosting also falls in this type of approach of manipulating the training data.

By generating an ensemble of classifiers and then combining them, the final rule produced is generally not in the space of rules that could be generated by the basic learning method. Thus, the ensemble method entertains a richer set of classifiers and so might be expected to have larger generalization error. One might argue that an ensemble method is just another learning method that considers a larger set of possible decision rules. It uses a set of training examples and produces a classifier. Yet, as we discussed for boosting, by suitably generating and combining the ensemble, performance better than any of the individual classifiers can often be obtained.

18.9   QUESTIONS

1. What is boosting? What is a "weak" learning rule? When might boosting be a useful method of improving the results of weak learning rules?

2. True or false: Boosting is best seen as a special case of a kernel method.

3. If $h_t(\overline{x})$ is the rule produced by boosting at round t, write the expression for the final classifier $H(\overline{x})$ after T rounds in terms of the $h_t(\overline{x})$ and the weights α_t.

4. How can boosting turn a weak classifier into a better classifier?

5. Philosophers sometimes suggest that there is a sense in which you are justified in believing something only if you believe it as the result of a reliable process of belief formation. What do you think they mean? Is there anything about statistical learning theory that might help explain what it is to be justified in this sense?

6. Consider the following method for using data to come up with a good classification rule: use the data to estimate the background probability distributions and then use these distributions to find a Bayes rule. How does this method compare with other learning methods discussed in this book?

18.10 REFERENCES

Kearns and Valiant (1988) considered the question of whether weak learners can be boosted to create a strong learner. Schapire (1990) provided a polynomial-time boosting algorithm, which was later improved by Freund (1995). The very popular and practical AdaBoost algorithm was introduced by Freund and Schapire (1997). There has been a great deal of work on analysis, refinements, extensions, and applications of boosting. The brief introduction by Schapire (1999), and survey papers by Freund and Schapire (1999), Schapire (2001), and Meir and Rätsch (2003) and references therein provide good entry points. Dietterich (2003) provides an overview of ensemble methods more generally.

Dietterich TG. Ensemble learning. In: The handbook of brain theory and neural networks. 2nd ed. Cambridge (MA): MIT Press; 2003.

Freund Y. Boosting a weak learning algorithm by majority. Inf Comput 1995;121(2): 256–285.

Freund Y, Schapire RE. A decision-theoretic generalization of on-line learning and an application to boosting. J Comput Syst Sci 1997;55(1):119–139.

Freund Y, Schapire RE. A short introduction to boosting. J Jpn Soc Artif Intel 1999;14(5):771–780. (In Japanese, translation by Naoki Abe.)

Kearns M, Valiant LG. Learning Boolean formulae or finite automata is as hard as factoring. Technical Report TR-14-88. Harvard University, Aiken Computation Laboratory; 1988.

Meir R, Rätsch G. An introduction to boosting and leveraging. In: Mendelson S, Smola A, editors. Advanced lectures on machine learning. LNCS. New York: Springer; 2003. pp. 119–184.

Schapire RE. The strength of weak learnability. Mach Learn 1990;5(2):197–227.

Schapire RE. A brief introduction to boosting. Proceedings of the Sixteenth International Joint Conference on Artificial Intelligence. Stockholm: 1999.

Schapire RE. The boosting approach to machine learning: an overview. MSRI Workshop on Nonlinear Estimation and Classification; Berkeley, CA. 2001.

Bibliography

Aizerman MA, Braverman EM, Rozonoer LI. Theoretical foundations of the potential function method in pattern recognition learning. Automat Remote Control 1964;25: 917–936.

Akaike H. An approximation to the density function. Ann Inst Stat Math 1954;6:127–132.

Akaike H. A new look at the statistical model identification. IEEE Trans Automat Control 1974;19:716–723.

Alon N, Ben-David S, Cesa-Bianchi N, Haussler D. Scale sensitive dimensions, uniform convergence, and learnability. Symposium on Foundations of Computer Science. Washington, D.C.: IEEE Computer Society Press; 1993.

Angluin D, Smith CH. Inductive inference: theory and methods. Comput Surv 1983; 15:237–269.

Anthony M, Bartlett PL. Neural network learning: theoretical foundations. Cambridge: Cambridge University Press; 1999.

Bartlett PL, Long PM, Williamson RC. Fat-shattering and the learnability of real-valued functions. J Comput Syst Sci 1996;52(3):434–452.

Bartlett P, Schölkopf B, Schuurmans D, Smola AJ, editors. Advances in large-margin classifiers. Cambridge (MA): MIT Press; 2000.

Bashkirov O, Braverman EM, Muchnik IE. Potential function algorithms for pattern recognition learning machines. Automat Remote Control 1964;25:692–695.

Bengson J, Moffett MA, Wright JC. The folk on knowing how. Phil Stud 2009;142:387–401.

Bertsekas D, Tsitsiklis J. Introduction to probability. 2nd ed. Belmont (MA): Athena Scientific; 2008.

Billingsly P. Probability and measure. 3rd ed. New York: Wiley Interscience; 1995.

Bishop C. Pattern recognition and machine learning. New York: Springer; 2006.

Blum M. A machine-independent theory of the complexity of recursive functions. J Assoc Comput Mach 1967;14:322–336.

Blum L, Blum M. Toward a mathematical theory of inductive inference. Inf Control 1975;28:125–155.

An Elementary Introduction to Statistical Learning Theory, First Edition.
Sanjeev Kulkarni and Gilbert Harman.
© 2011 John Wiley & Sons, Inc. Published 2011 by John Wiley & Sons, Inc.

Blumer A, Ehrenfeucht A, Haussler D, Warmuth M. Learnability and the Vapnik-Chervonenkis dimension. J ACM 1989;36(4):929–965.

Bongard M. Pattern recognition. Washington (DC): Spartan Books; 1970.

Boser B, Guyon I, Vapnik VN. A training algorithm for optimal margin classifiers. Proceedings of the 5th Annual ACM Workshop on Computational Learning Theory. New York: Association for Computing Machinery; 1992. pp. 144–152.

Braverman EM. The method of potential functions. Automat Remote Control 1965;26:2130–2138.

Burges CJC. A tutorial on support vector machines for pattern recognition. Data Mining Knowl Discov 1998;2:121–167.

Chung KL. A course in probability theory. 2nd ed. Amsterdam: Academic Press; 2000.

Corder GW, Foreman DI. Nonparametric statistics for non-statisticians: a step-by-step approach. Hoboken (NJ): Wiley; 2009.

Corfield D, Schölkopf B, Vapnik V. Falsificationism and statistical learning theory: comparing the popper and Vapnik-Chervonenkis dimensions. J Gen Philos Sci. 2009;40:51–58.

Cortes C, Vapnik VN. Support vector networks. Mach Learn 1995;20:1–25.

Cover TM. Estimation by the nearest neighbor rule. IEEE Trans Inf Theory 1968;IT-14:50–55.

Cover TM, Hart PE. Nearest neighbor pattern classification. IEEE Trans Inf Theory 1967;13(1):21–27.

Cristianini N, Shawe-Taylor J. An introduction to support vector machines. Cambridge University Press; 2000.

Cybenko G. Approximations by superpositions of sigmoidal functions. Math Control Signals Syst 1989;2(4):303–314.

Dasarathy BV, editor. Nearest Neighbor (NN) Norms: NN pattern classification techniques. Washington (DC): IEEE Computer Society; 1991.

Della Rocca M. Two spheres, twenty spheres, and the identity of indiscernibles. Pac Philos Q 2005;86:480–492.

Descartes R. Meditationes de Prima Philosophia. Paris; 1641.

Devijver PR, Kittler J. Pattern recognition: a statistical approach. Englewood Cliffs (NJ): Prentice-Hall; 1982.

Devroye L, Györfi L, Lugosi G. A probabilistic theory of pattern recognition. New York: Springer Verlag; 1996.

Dewey J. Logic: the theory of inquiry. New York: Holt; 1938.

Dietterich TG. Ensemble learning. The handbook of brain theory and neural networks. 2nd ed. Cambridge (MA): MIT Press; 2003

Duda RO, Hart PE, Pattern classification and Scene Analysis. New York: Wiley; 1973.

Duda RO, Hart PE, Stork DG. Pattern classification. 2nd ed. New York: Wiley; 2001.

Dudley RM. Real analysis and probability. 2nd ed. Cambridge: Cambridge University Press; 2002.

Dworkin R. Law's empire. Cambridge (MA): Harvard UP; 1986.

Everitt BS. Chance rules: an informal guide to probability, risk, and statistics. New York: Copernicus, Springer-Verlag; 1999.

Feller W. Volume 1, An introduction to probability theory and its applications. 3rd ed. New York: Wiley; 1968.

Fix E, Hodges JL. Discriminatory analysis: nonparametric discrimination: consistency properties. USAF Sch Aviat Med 1951;4:261–279.

Fix E, Hodges JL. Discriminatory analysis: nonparametric discrimination: small sample performance. USAF Sch Aviat Med 1952;11:280–322.

Forster MR, Sober E. How to tell when simpler, more unified, or less Ad Hoc theories will provide more accurate predictions. Br J Philos Sci 1994;45:1–35.

Freund Y. Boosting a weak learning algorithm by majority. Inf Comput 1995;121(2): 256–285.

Freund Y, Schapire RE. A decision-theoretic generalization of on-line learning and an application to boosting. J Comput Syst Sci 1997;55(1):119–139.

Freund Y, Schapire RE. A short introduction to boosting. J Jpn Soc Artif Intell 1999;14(5):771–780. (In Japanese, translation by Naoki Abe.)

Fukunaga K. Introduction to statistical pattern recognition. 2nd ed. San Diego (CA): Academic Press; 1990.

Gigerenzer G, et al., Simple heuristics that make us smart. Oxford: Oxford University Press; 1999.

Gold EM. Language identification in the limit. Inf Control 1967;10:447–474.

Goodman N. Fact, fiction, and forecast. 2nd ed. Indianapolis (IN): Bobbs-Merrill; 1965.

Györfi L, editor. Principles of nonparametric learning. New York: Springer; 2002.

Györfi L, Härdle W, Sarda P, Vieu P. Nonparametric curve estimation from time series. Lecture notes in statistics. Berlin: Springer-Verlag; 1989.

Györfi L, Kohler M, Krzyzak A, Walk H. A distribution-free theory of nonparametric regression. Springer; 2002.

Hacking I. The logic of statistical inference. Cambridge: Cambridge University Press; 1965.

Hacking I. The emergence of probability. Cambridge: Cambridge University Press; 1984.

Hájek A. What conditional probability could not be. Synthese 2003;137:273–323.

Hardle W. Applied nonparametric regression. Cambridge University Press; 1992.

Harman G. Change in view: principles of reasoning. Cambridge (MA): MIT Press; 1986.

Harman G. Simplicity as a pragmatic criterion for deciding what hypotheses to take seriously. In: Harman G, editor. Reasoning, meaning, and mind. Oxford: Oxford University Press; 1999.

Harman G, Kulkarni S. Reliable reasoning: induction and statistical learning theory. Cambridge (MA): MIT Press; 2007.

Hastie T, Tibshirani R, Friedman J. The elements of statistical learning: data mining, inference, and prediction. 2nd ed. Springer; 2009.

Haussler D. Decision theoretic generalizations of the PAC model for neural net and other learning applications. Inf Comput 1992;100:78–150.

Haykin S. Neural networks: a comprehensive foundation. New York: Macmillan Publishing Company; 1994.

Ho YC, Agrawala A. On pattern classification algorithms: introduction and survey. Proc IEEE 1968;56:2101–2114.

Holst M, Irle A. Nearest neighbor classification with dependent training sequences. Ann Stat 2001;29(5):1424–1442.

Kearns MJ, Schapire RE. Efficient distribution-free learning of probabilistic concepts. J Comput Syst Sci 1994;48(3):464–497.

Kearns M, Valiant LG. Learning Boolean formulae or finite automata is as hard as factoring Technical Report TR-14-88. Harvard University, Aiken Computation Laboratory; 1988.

Kearns MJ, Vazirani UV. An introduction to computational learning theory. Cambridge (MA): MIT Press; 1994.

Kolmogorov AN. Grundbegriffe der wahrscheinlichkeitrechnung. Ergebnisse der mathematik; translated as foundations of probability. New York: Chelsea Publishing Company; 1933; 1950.

Kugel P. Induction, pure and simple. Inf Control 1977;35:276–336.

Kulkarni SR, Lugosi G, Venkatesh S. Learning pattern classification—A survey. IEEE Trans Inf Theory 1998;44(6):2178–2206.

Lozano A, Kulkarni SR, Schapire R. Convergence and consistency of regularized boosting algorithms with stationary β-mixing observations. Volume 18, Advances in neural information processing systems; 2006.

Ludlow P. Simplicity and generative grammar. In Stainton R, Murasugi K, editors. Philosophy and linguistics. Boulder (CO): Westview Press; 1998.

Mach E. The science of mechanics. 6th ed. Chicago: Open Court; 1960.

McCulloch WS, Pitts W. A logical calculus of the ideas immanent in nervous activity. Bull Math Biophys 1943;5:115–133.

Meir R, Rätsch G. An introduction to boosting and leveraging. In: Mendelson S, Smola A, editors. Advanced lectures on machine learning, LNCS. New York: Springer; 2003. pp. 119–184.

Minsky M, Papert S. Perceptrons. Expanded edition. Cambridge (MA): MIT Press; 1988.

Mitchell TM. Machine learning. New York: McGraw-Hill; 1997. pp. 226–229.

Nadaraya EA. On estimating regression. Theory Probab Appl 1964;9:141–142.

Nadaraya EA. Remarks on nonparametric estimates for density functions and regression curves. Theory Probab Appl 1970;15:134–137.

Nadaraya EA. Nonparametric estimation of probability densities and regression curves. Dordrecht: Kluwer; 1989.

Nagy G. State of the art in pattern recognition. Proc IEEE 1968;56:836–862.

Nilsson NJ. Learning machines. New York: McGraw-Hill; 1965.

Nozick R. Simplicity as fall-out. In: Cauman L, Levi I, Parsons C, editors. How many questions: essays in honor of sydney morgenbesser. Indianapolis: Hackett Publishing; 1983. pp. 105–119.

Parzen E. On the estimation of a probability density function and the mode. Ann Math Stat 1962;33:1065–1076.

Peirce CS. In: Hartshorne C, Weiss P, Burks A, editors. Collected papers of Charles Sanders Peirce, 8 vols. Cambridge: Harvard University Press; 1931–58.

Poldrak RA, Halchenko YO, Hanson SJ. Decoding the large-scale structure of brain function by classifying mental states across individuals. Psychol Sci 2009;20(11):1364–1372.

Pollard D. Convergence of stochastic processes. New York: Springer; 1984.

Popper K. Objective knowledge: an evolutionary approach. Oxford: Clarendon Press; 1979.

Popper K. The logic of scientific discovery. London: Routledge; 2002.

Putnam H. "Degree of confirmation" and inductive logic. In: Schillp A, editor. The philosophy of Rudolph Carnap. LaSalle (IL): Open Court; 1963.

Rabinowitz L. Elementary probability with applications. Wellesley (MA): AK Peters; 2004.

Rawls J. A theory of justice. Cambridge (MA): Harvard University Press; 1971.

Rissanen J. Volume 15, Stochastic complexity in statistical inquiry, Singapore: World Scientific, Series in Computer Science; 1989.

Rosenblatt M. Remarks on some nonparametric estimates of a density function. Ann Math Stat 1956;27:832–837.

Rosenblatt F. The perceptron: a probabilistic model for information storage and organization in the brain. Psychol Rev 1958;65:386–408.

Rosenblatt F. Perceptron simulation experiments. Proc Inst Radio Eng 1960;48:301–309.

Rosenblatt F. Principles of neurodynamics. Washington (DC): Spartan Books; 1962.

Ross SM. A first course in probability. 8th ed. Upper Saddle River (NJ): Prentice-Hall; 2009.

Rumelhart DE, McClelland JL, editors. Volume 1, Parallel distributed processing: explorations in the microstructure of cognition. Cambridge (MA): MIT Press; 1986.

Rumelhart DE, Hinton GE, McClelland JL. Learning representations by back-propagating errors. Nature 1986a;323:533–536.

Rumelhart DE, Hinton GE, McClelland JL. Learning internal representations by error propagation. In: Rumelhart DE, McClelland JL, editors. Volume 1, Parallel distributed processing. Cambridge (MA): MIT Press; 1986b, Chapter 8.

Russell S, Norvig P. Artificial intelligence: a modern approach. Learning from examples. Upper Saddle River (NJ): Prentice-Hall; 2010. pp. 645–767, Chapter 18.

Schalkoff RJ. Pattern recognition: statistical, structural, and neural approaches. New York: Wiley; 1992.

Schapire RE. The strength of weak learnability. Mach Learn 1990;5(2):197–227.

Schapire RE. A brief introduction to boosting. Proceedings of the Sixteenth International Joint Conference on Artificial Intelligence. Stockholm; 1999.

Schapire RE. The boosting approach to machine learning: an overview. MSRI Workshop on Nonlinear Estimation and Classification; 2001 Mar; Berkeley (CA).

Schölkopf B, Burges CJC, Smola AJ, editors. Advances in kernel methods: support vector learning. Cambridge (MA): MIT Press; 1999.

Schölkopf B, Smola AJ. Learning with kernels: support vector machines, regularization, optimization, and beyond (Adaptive Computation and Machine Learning). Cambridge (MA): MIT Press; 2001.

Scott DW. Multivariate density estimation: theory, practice and visualization. New York: Wiley; 1992.

Shawe-Taylor J, Cristianini N. Kernel methods for pattern analysis. Cambridge: Cambridge University Press; 2004.

Silverman BW. Density estimation. London: Chapman and Hall; 1986.

Sober E. Simplicity. Oxford: Oxford University Press; 1975.

Sober E. Reconstructing the past. Cambridge (MA): MIT Press; 1988.

Sober E. Let's razor Occam's razor. In: Knowles D, editor. Explanation and its limits. Cambridge: Cambridge University Press; 1990.

Solomonoff RJ. A formal theory of inductive inference. Inf Control 1964;7:1–22, 224–254.

Stalker D. Grue: the new riddle of induction. Peru (IL): Open Court; 1994.

Stanley J, Williamson T. Knowing how. J Philos 2001;98:411–444.

Steinwart I, Hush D, Scovel C. Learning from dependent observations. J Multivariate Anal 2009;100:175–194.

Stich SP. Moral philosophy and mental representation. In: Hechter M, Nadel L, Michod R, editors. The origin of values. Hawthorne (NY): Aldine de Gruyter; 1993. pp. 215–228.

Stone CJ. Consistent nonparametric regression. Ann Stat 1977;5:595–645.

Theodoridis S, Koutroumbas K. Pattern recognition. 4th ed. Amsterdam: Academic Press; 2008.

Theodoridis S, Pikrakis A, Koutroumbas K, Cavouras D. Introduction to pattern recognition: a matlab approach. Amsterdam: Academic Press; 2010.

Tijms H. Understanding probability: chance rules in everyday life. Cambridge: Cambridge University Press; 2007.

Turney PD. Inductive inference and stability [PhD dissertation]. Department of Philosophy, University of Toronto; 1988.

Valiant LG. The complexity of enumeration and reliability problems. SIAM J Comput 1979;8:410–421.

Valiant LG. A theory of the learnable. Commun ACM 1984;27(11):1134–1142.

Vapnik VN. Estimation of dependencies based on empirical data. New York: Springer-Verlag; 1982.

Vapnik VN. The nature of statistical learning theory. New York: Springer; 1991.

Vapnik VN. The nature of statistical learning theory. New York: Springer-Verlag; 1996.

Vapnik VN. The nature of statistical learning theory. New York: Springer; 1999.

Vapnik V. Statistical learning theory. New York: Wiley-Interscience; 1998.

Vapnik VN, Chervonenkis A. On the uniform convergence of relative frequencies of events to their probabilities. Theory Probab Appl 1971;16(2):264–280.

Vickers J. The Problem of Induction, in The Stanford Encyclopedia of Philosophy; 2010, http://plato.stanford.edu/entries/induction-problem/.

Vidyasagar M. A theory of learning and generalization. London: Springer-Verlag; 1997.

Vidyasagar M. Convergence of empirical means with α-mixing input sequences, and an application to PAC learning. Proceedings of the 44th IEEE Conference on Decision and Control, and the European Control Conference; 2005. pp. 560–565.

Watanabe MS. Knowing and guessing. New York: Wiley; 1969.

Watson GS. Smooth regression analysis. Sankhya Ser A 1964;26:359–372.

Werbos PJ. Beyond regression: new tools for prediction and analysis in the behavioral sciences [PhD thesis]. Cambridge (MA); Harvard University; 1974.

Widrow B, Generalization and information storage in networks of adaline 'neurons''. In: Yovitz MC, editor. Self-organizing systems. Wahington (DC): Spartan; 1962. pp. 435–461.

Widrow B, Hoff ME. Adaptive switching circuits. IRE WESCON Convention Record; 1960; 4:96–104.

Winograd T, Flores F. Understanding computers and cognition. Norwood (NJ): Ablex; 1986.

Zellner A, Keuzenkamp H, McAleer M,editors. Simplicity, inference and econometric modelling. Cambridge: Cambridge University Press; 2002.

Author Index

Agrawala, A., 8, 42
Aizerman, M. A., 84, 85
Akaike, H., 84, 85, 139, 143
Alon, N., 158
Anguluin, D. C., 170, 171
Anthony, M., 124, 136

Banach, S., 31
Bartlett, P. L., 124, 136, 158, 186
Bashkirov, O., 84, 85
Bayes, T., 53
Ben-David, S., 158
Bengson, J., 62, 64
Bernoulli, J., 22
Billingsly, P., 33
Bishop, C., 8
Blum, L., 170, 171
Blum, M., 170, 171
Blumer, A., 136
Bongard, M., 8, 41, 42
Borel, E., 32
Boser, B., 186
Braverman E. M., 84, 85
Burges, C. J. C., 186
Burks, A., 171

Cardano, G., 22
Cavouras, D., 8
Cesa-Bianchi, N., 158
Chervonenkis, A., 123, 124, 128, 136
Cristianini, N., 85, 186
Chung, K. L., 33

Church, A., 184
Corder, G. W., 149
Corfield, D., 136
Cortes, C., 186
Cover, T. M., 73, 158
Cybenko, G., 158

Dasarathy, B., V., 73
Della Rocca, M., 124
de Moivre, A., 22
Descartes, R., 95, 97
Devijver, P. R., 8, 41, 42
Devroye, L., 8, 41, 42, 53, 54, 64, 73, 85, 124, 136, 143
Dewey, J., 170, 171
Dieterich, T. G., 196
Duda, R. O., 8, 41, 42, 54, 64, 73, 97, 115
Dudley, R. M., 33
Dworkin, R., 73

Ehrenfeucht, A., 136
Einstein, A., 169
Everitt, B. S., 22, 53, 54

Feller, W., 22, 30, 33
Fermat, P., 22
Fisher, R., 22
Fix, E., 73
Flores, F., 73
Foreman, D. I., 149
Forster, M. R., 171
Fraenkel, A.H., 32

An Elementary Introduction to Statistical Learning Theory, First Edition.
Sanjeev Kulkarni and Gilbert Harman.
© 2011 John Wiley & Sons, Inc. Published 2011 by John Wiley & Sons, Inc.

Subject Index

An Elementary Introduction to Statistical Learning Theory, First Edition.
Sanjeev Kulkarni and Gilbert Harman.
© 2011 John Wiley & Sons, Inc. Published 2011 by John Wiley & Sons, Inc.

WILEY SERIES IN PROBABILITY AND STATISTICS
ESTABLISHED BY WALTER A. SHEWHART AND SAMUEL S. WILKS

Editors: *David J. Balding, Noel A. C. Cressie, Garrett M. Fitzmaurice,*
Harvey Goldstein, Iain M. Johnstone, Geert Molenberghs,
David W. Scott, Adrian F. M. Smith, Ruey S. Tsay, Sanford Weisberg
Editors Emeriti: *Vic Barnett, J. Stuart Hunter, Joseph B. Kadane,*
Jozef L. Teugels

The *Wiley Series in Probability and Statistics* is well established and authoritative. It covers many topics of current research interest in both pure and applied statistics and probability theory. Written by leading statisticians and institutions, the titles span both state-of-the-art developments in the field and classical methods.

Reflecting the wide range of current research in statistics, the series encompasses applied, methodological and theoretical statistics, ranging from applications and new techniques made possible by advances in computerized practice to rigorous treatment of theoretical approaches.

This series provides essential and invaluable reading for all statisticians, whether in academia, industry, government, or research.

† ABRAHAM and LEDOLTER · Statistical Methods for Forecasting
  AGRESTI · Analysis of Ordinal Categorical Data, *Second Edition*
  AGRESTI · An Introduction to Categorical Data Analysis, *Second Edition*
  AGRESTI · Categorical Data Analysis, *Second Edition*
  ALTMAN, GILL, and McDONALD · Numerical Issues in Statistical Computing for the
     Social Scientist
  AMARATUNGA and CABRERA · Exploration and Analysis of DNA Microarray and
     Protein Array Data
  ANDĚL · Mathematics of Chance
  ANDERSON · An Introduction to Multivariate Statistical Analysis, *Third Edition*
* ANDERSON · The Statistical Analysis of Time Series
  ANDERSON, AUQUIER, HAUCK, OAKES, VANDAELE, and WEISBERG ·
     Statistical Methods for Comparative Studies
  ANDERSON and LOYNES · The Teaching of Practical Statistics
  ARMITAGE and DAVID (editors) · Advances in Biometry
  ARNOLD, BALAKRISHNAN, and NAGARAJA · Records
* ARTHANARI and DODGE · Mathematical Programming in Statistics
* BAILEY · The Elements of Stochastic Processes with Applications to the Natural
     Sciences
  BALAKRISHNAN and KOUTRAS · Runs and Scans with Applications
  BALAKRISHNAN and NG · Precedence-Type Tests and Applications
  BARNETT · Comparative Statistical Inference, *Third Edition*
  BARNETT · Environmental Statistics
  BARNETT and LEWIS · Outliers in Statistical Data, *Third Edition*
  BARTOSZYNSKI and NIEWIADOMSKA-BUGAJ · Probability and Statistical Inference
  BASILEVSKY · Statistical Factor Analysis and Related Methods: Theory and
     Applications
  BASU and RIGDON · Statistical Methods for the Reliability of Repairable Systems
  BATES and WATTS · Nonlinear Regression Analysis and Its Applications

*Now available in a lower priced paperback edition in the Wiley Classics Library.
†Now available in a lower priced paperback edition in the Wiley–Interscience Paperback Series.

BECHHOFER, SANTNER, and GOLDSMAN · Design and Analysis of Experiments for Statistical Selection, Screening, and Multiple Comparisons

BELSLEY · Conditioning Diagnostics: Collinearity and Weak Data in Regression

† BELSLEY, KUH, and WELSCH · Regression Diagnostics: Identifying Influential Data and Sources of Collinearity

BENDAT and PIERSOL · Random Data: Analysis and Measurement Procedures, *Fourth Edition*

BERRY, CHALONER, and GEWEKE · Bayesian Analysis in Statistics and Econometrics: Essays in Honor of Arnold Zellner

BERNARDO and SMITH · Bayesian Theory

BHAT and MILLER · Elements of Applied Stochastic Processes, *Third Edition*

BHATTACHARYA and WAYMIRE · Stochastic Processes with Applications

BILLINGSLEY · Convergence of Probability Measures, *Second Edition*

BILLINGSLEY · Probability and Measure, *Third Edition*

BIRKES and DODGE · Alternative Methods of Regression

BISGAARD and KULAHCI · Time Series Analysis and Forecasting by Example

BISWAS, DATTA, FINE, and SEGAL · Statistical Advances in the Biomedical Sciences: Clinical Trials, Epidemiology, Survival Analysis, and Bioinformatics

BLISCHKE AND MURTHY (editors) · Case Studies in Reliability and Maintenance

BLISCHKE AND MURTHY · Reliability: Modeling, Prediction, and Optimization

BLOOMFIELD · Fourier Analysis of Time Series: An Introduction, *Second Edition*

BOLLEN · Structural Equations with Latent Variables

BOLLEN and CURRAN · Latent Curve Models: A Structural Equation Perspective

BOROVKOV · Ergodicity and Stability of Stochastic Processes

BOULEAU · Numerical Methods for Stochastic Processes

BOX · Bayesian Inference in Statistical Analysis

BOX · R. A. Fisher, the Life of a Scientist

BOX and DRAPER · Response Surfaces, Mixtures, and Ridge Analyses, *Second Edition*

* BOX and DRAPER · Evolutionary Operation: A Statistical Method for Process Improvement

BOX and FRIENDS · Improving Almost Anything, *Revised Edition*

BOX, HUNTER, and HUNTER · Statistics for Experimenters: Design, Innovation, and Discovery, *Second Editon*

BOX, JENKINS, and REINSEL · Time Series Analysis: Forcasting and Control, *Fourth Edition*

BOX, LUCEÑO, and PANIAGUA-QUIÑONES · Statistical Control by Monitoring and Adjustment, *Second Edition*

BRANDIMARTE · Numerical Methods in Finance: A MATLAB-Based Introduction

† BROWN and HOLLANDER · Statistics: A Biomedical Introduction

BRUNNER, DOMHOF, and LANGER · Nonparametric Analysis of Longitudinal Data in Factorial Experiments

BUCKLEW · Large Deviation Techniques in Decision, Simulation, and Estimation

CAIROLI and DALANG · Sequential Stochastic Optimization

CASTILLO, HADI, BALAKRISHNAN, and SARABIA · Extreme Value and Related Models with Applications in Engineering and Science

CHAN · Time Series: Applications to Finance with R and S-Plus®, *Second Edition*

CHARALAMBIDES · Combinatorial Methods in Discrete Distributions

CHATTERJEE and HADI · Regression Analysis by Example, *Fourth Edition*

CHATTERJEE and HADI · Sensitivity Analysis in Linear Regression

CHERNICK · Bootstrap Methods: A Guide for Practitioners and Researchers, *Second Edition*

CHERNICK and FRIIS · Introductory Biostatistics for the Health Sciences

CHILÈS and DELFINER · Geostatistics: Modeling Spatial Uncertainty

*Now available in a lower priced paperback edition in the Wiley Classics Library.
†Now available in a lower priced paperback edition in the Wiley–Interscience Paperback Series.

*Now available in a lower priced paperback edition in the Wiley Classics Library.
†Now available in a lower priced paperback edition in the Wiley–Interscience Paperback Series.

*Now available in a lower priced paperback edition in the Wiley Classics Library.
†Now available in a lower priced paperback edition in the Wiley–Interscience Paperback Series.

*Now available in a lower priced paperback edition in the Wiley Classics Library.
†Now available in a lower priced paperback edition in the Wiley–Interscience Paperback Series.

*Now available in a lower priced paperback edition in the Wiley Classics Library.
†Now available in a lower priced paperback edition in the Wiley–Interscience Paperback Series.

*Now available in a lower priced paperback edition in the Wiley Classics Library.
†Now available in a lower priced paperback edition in the Wiley–Interscience Paperback Series.

TAMHANE · Statistical Analysis of Designed Experiments: Theory and Applications
TANAKA · Time Series Analysis: Nonstationary and Noninvertible Distribution Theory
THOMPSON · Empirical Model Building
THOMPSON · Sampling, *Second Edition*
THOMPSON · Simulation: A Modeler's Approach
THOMPSON and SEBER · Adaptive Sampling
THOMPSON, WILLIAMS, and FINDLAY · Models for Investors in Real World Markets
TIAO, BISGAARD, HILL, PEÑA, and STIGLER (editors) · Box on Quality and
  Discovery: with Design, Control, and Robustness
TIERNEY · LISP-STAT: An Object-Oriented Environment for Statistical Computing
  and Dynamic Graphics
TSAY · Analysis of Financial Time Series, *Third Edition*
UPTON and FINGLETON · Spatial Data Analysis by Example, Volume II:
  Categorical and Directional Data
† VAN BELLE · Statistical Rules of Thumb, *Second Edition*
VAN BELLE, FISHER, HEAGERTY, and LUMLEY · Biostatistics: A Methodology for
  the Health Sciences, *Second Edition*
VESTRUP · The Theory of Measures and Integration
VIDAKOVIC · Statistical Modeling by Wavelets
VINOD and REAGLE · Preparing for the Worst: Incorporating Downside Risk in Stock
  Market Investments
WALLER and GOTWAY · Applied Spatial Statistics for Public Health Data
WEERAHANDI · Generalized Inference in Repeated Measures: Exact Methods in
    MANOVA and Mixed Models
WEISBERG · Applied Linear Regression, *Third Edition*
WEISBERG · Bias and Causation: Models and Judgment for Valid Comparisons
WELSH · Aspects of Statistical Inference
WESTFALL and YOUNG · Resampling-Based Multiple Testing: Examples and
  Methods for *p*-Value Adjustment
WHITTAKER · Graphical Models in Applied Multivariate Statistics
WINKER · Optimization Heuristics in Economics: Applications of Threshold Accepting
WONNACOTT and WONNACOTT · Econometrics, *Second Edition*
WOODING · Planning Pharmaceutical Clinical Trials: Basic Statistical Principles
WOODWORTH · Biostatistics: A Bayesian Introduction
WOOLSON and CLARKE · Statistical Methods for the Analysis of Biomedical Data,
  Second Edition
WU and HAMADA · Experiments: Planning, Analysis, and Parameter Design
  Optimization, *Second Edition*
WU and ZHANG · Nonparametric Regression Methods for Longitudinal Data Analysis
YANG · The Construction Theory of Denumerable Markov Processes
YOUNG, VALERO-MORA, and FRIENDLY · Visual Statistics: Seeing Data with
  Dynamic Interactive Graphics
ZACKS · Stage-Wise Adaptive Designs
ZELTERMAN · Discrete Distributions—Applications in the Health Sciences
* ZELLNER · An Introduction to Bayesian Inference in Econometrics
ZHOU, OBUCHOWSKI, and McCLISH · Statistical Methods in Diagnostic Medicine,
  Second Edition

Printed in the United States
By Bookmasters